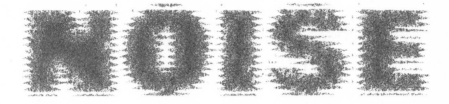

NOISE

BART KOSKO, Ph.D.

VIKING

VIKING

Published by the Penguin Group

Penguin Group (USA) Inc., 375 Hudson Street, New York, New York 10014, U.S.A.

Penguin Group (Canada), 90 Eglinton Avenue East, Suite 700, Toronto, Ontario, Canada M4P 2Y3 (a division of Pearson Penguin Canada Inc.)

Penguin Books Ltd, 80 Strand, London WC2R 0RL, England

Penguin Ireland, 25 St. Stephen's Green, Dublin 2, Ireland (a division of Penguin Books Ltd)

Penguin Books Australia Ltd, 250 Camberwell Road, Camberwell, Victoria 3124, Australia (a division of Pearson Australia Group Pty Ltd)

Penguin Books India Pvt Ltd, 11 Community Centre, Panchsheel Park, New Delhi–110 017, India

Penguin Group (NZ), Cnr Airborne and Rosedale Roads, Albany, Auckland 1310, New Zealand (a division of Pearson New Zealand Ltd)

Penguin Books (South Africa) (Pty) Ltd, 24 Sturdee Avenue, Rosebank, Johannesburg 2196, South Africa

Penguin Books Ltd, Registered Offices:
80 Strand, London WC2R 0RL, England

First published in 2006 by Viking Penguin,
a member of Penguin Group (USA) Inc.

10   9   8   7   6   5   4   3   2   1

Photograph of Hedy Lamarr on page 139 © John Springer Collection/CORBIS

ISBN 0-670-03495-9

Printed in the United States of America
Set in Berthold Baskerville Book with Cathode and Tube
Designed by Daniel Lagin

For Dorothy Kosko (1922–2000)

Good neighbors keep their noise to themselves.
　　　−motto of the Noise Pollution Clearinghouse

ACKNOWLEDGMENTS

The author thanks editor Rick Kot for his exceptional editorial judgment and his equally exceptional patience.

CONTENTS

I hate noise. I hate the roar of car traffic and leaf blowers and airplane overflights. I hate the screech of car alarms and police sirens and the public speeches that too many of my fellow citizens make into their private cell phones. I dislike any unwanted signal that impinges on my humble sense organs. This is especially so when that signal interferes with other signals of interest—such as music or a friend's voice or when it simply intrudes on the sound of silence. The same holds for the cyber-equivalent noise that arrives as spam in an e-mail mailbox or that flashes on the side of the computer screen as the latest product advertisement. They are all noise because they are all unwanted signals.

I am not alone in this dislike of noise. Polls and anecdotal data show that city dwellers tend to hate noise as much as they hate congestion or crime or any other city blight. New York City is a case in point because it is one of the noisiest cities in the world. New York mayor Michael Bloomberg began a legal war on noise pollution in 2004 with his overhaul of the city's outdated noise code. Such legal challenges to noise pollution will only continue as the problem grows. Even those of us who favor limited government have to grapple with this growing market failure as we impose ever more unwanted noise pollution on one another. Some form of local or state or even federal government regulation may be the only economically efficient solution in many cases because of the high transactions costs involved in getting others to tone it down. Anti-noise-pollution activism may become a new political force in the ever noisier digital age—much as anti-environmental-pollution activism did in the earlier industrial age.

There is good reason to hate noise in the city or in the bedroom or anywhere else. Noise causes hearing loss in millions of people. It causes stress and sleep loss. Those maladies in turn promote more bad health effects and an overall decreased sense of well-being. Recent studies have shown that noise even increases the risk of heart attack and high blood pressure. Noise pollution also disturbs the low-decibel environment in ways that we are just beginning to understand. Underwater sonar noise appears to make humpback whales sing longer to communicate with one another. A Dutch study found that city noise makes the small bird known as the great tit sing at ever higher frequencies to communicate. Sustained noise energy may well produce other effects on the sensory and brain tissue of these and thousands of other species.

The future promises ever more noise. The modern world tends to become an ever louder and higher-decibel place that differs in kind from the low-decibel environments in which our hominid ancestors evolved over millions of years. More cars and gadgets and machinery produce this growing rumble of noise. Hospital noise is a case in point. The World Health Organization recommends an average noise level of about 35 decibels in patient areas. But a recent Johns Hopkins study found that the average noise level in hospitals rose from 57 decibels in 1965 to 72 decibels in 2005 with an even greater percentage increase in hospital evening noise. Noise pollution will only grow as the global economy expands.

I have another reason to dislike noise: I wage war on noise every day as part of my work as a scientist and engineer. We try to maximize signal-to-noise ratios. We try to filter noise out of measurements of sounds or images or anything else that conveys information from the world around us. We code the transmission of digital messages with extra 0s and 1s to defeat line noise and burst noise and any other form of interference. We design sophisticated algorithms to track noise and then cancel it in headphones or in a sonogram. Some of us even teach classes on how to defeat this nemesis of the digital age. Such action further conditions our anti-noise reflexes.

So it came as a surprise to find that sometimes noise is not an enemy. There had been a few engineering reports of the benefits of noise or "dithering" as early as the 1960s. I came upon the phenomenon myself in the 1980s while working with neural networks called associative memories. These networks consisted of simple on-off neurons that could learn and store patterns. The networks could later recall those patterns if a user showed the network only a small or noisy portion of them—just as we can recognize a tune after we hear only a small portion of it. Adding noise would sometimes smooth out the learning process or help stabilize a feedback network much as random raindrops can help calm a rough sea. Other scientists and engineers found and published more noise-is-good results and soon a new field of sorts called *stochastic resonance* started to emerge in the 1990s.

The sociology of the growing noise movement resembled the rise of chaos theory in the early 1980s and the rise of fuzzy logic in the late 1980s. Papers on noise benefits appeared in disparate journals and conferences because there was no journal or technical society dedicated to noise benefits. That remains true today with some small but important exceptions. Noise appeared to be a type of garbage-bin technical property much as chaos and fuzz first appeared to earlier audiences of scientists and engineers. Today fuzzy logic enhances the performance of thousands of consumer products and industrial systems by storing and reasoning with commonsense

rules in computer chips or software. And chaos theory describes behavior that ranges from minuscule quantum fluctuations to heart
arrhythmias and to the secure transmission of information in fiber-
optical networks. Scientists have likewise now found hundreds of
noise benefits because they have started looking for them–even
though the noise benefits have been there all along.

This book looks at noise both as something bad and as something good. That fuzzy view of noise involves forays into fields such
as law and economics and neural physiology as well as into the more
formal noise haunts in science and engineering. The result is only an
introduction to this vast and expanding subject and by no means a
comprehensive treatment of it. The endnotes point the way to some
of the pertinent technical literature on noise. Space demands require
that they too are far from comprehensive.

The common goal in this noise inquiry is to see more noise patterns as signals whether or not we like those signals. This change in
perspective involves a step up in understanding–as when we first see
that many of our own favorite signals strike others as noise. Insight
grows along with an implied signal-to-noise ratio when more noise
counts as signal. The shift from noise to signal starts at birth and
continues throughout life. Thousands of sensory signals impinge on
a newborn child but the child sees and hears and feels and tastes
many of these as noise. An adult perceives most of the same signals
as legitimate signals unto themselves. A physician or other expert
sees even more signal structure in the same swath of experience.
And so it goes up the ladder of insight–to God all is signal.

CHAPTER 1

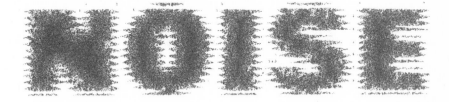

Noise signals are unwanted signals that are always present in a transmission system.

—John R. Pierce
Signals: The Telephone and Beyond

Noise is unwanted sound. It is derived from the Latin word "nausea" meaning seasickness. Noise is among the most pervasive pollutants today. Noise from road traffic, jet planes, jet skis, garbage trucks, construction equipment, manufacturing processes, lawn mowers, leaf blowers, and boom boxes, to name a few, are among the unwanted sounds that are routinely broadcast into the air.

—Noise Pollution Clearinghouse Web site

The term "environmental noise" means the intensity, duration, and the character of sounds from all sources.

—United States Code, Title 42, Section 4902

Sounds can be divided into noise and music. A very musical sound can be regarded as noise if it distracts the listener from some other more important sound. The beautiful performance of a Debussy prelude in the room next door may be regarded as intolerable noise by someone trying to write program notes on a Beethoven sonata.

–Charles Taylor

Sound

If the channel is noisy it is not in general possible to reconstruct the original message or the transmitted signal with certainty by any operation on the received signal. There are ways, however, of transmitting the information which are optimal in combating noise.

–Claude E. Shannon (1916–2001)

"A Mathematical Theory of Communication"

Bell System Technical Journal, July 1948

In most cases the worst enemy is the noise brought in by the environment that surrounds the equipment. The environment includes association with power lines, earth, equipment, cables, and external transmitters. The constant battle is how to make equipment work well in each and every situation.

–Ralph Morrison

Noise and Other Interfering Signals

Engineers have sought to minimize the effects of noise in electronic circuits and communication systems. But recent research has established that noise can play a constructive role in the detection of weak periodic signals.

–Kurt Wiesenfeld and Frank Moss

"Stochastic Resonance and the Benefits of Noise:
From Ice Ages to Crayfish and SQUIDs"

Nature, vol. 373, 5 January 1995

Stochastic resonance simply stands for a new paradigm wherein noise represents a useful tool rather than a nuisance.

> –Luca Gammaitoni, Peter Hänggi, Peter Jung,
> and Fabio Marchesoni
> "Stochastic Resonance"
> *Reviews of Modern Physics*, vol. 70, no. 1, January 1998

Uncovering the mysteries of natural phenomena that were formerly someone else's "noise" is a recurring theme in science.

> –Alfred Bedard Jr. and Thomas Georges
> "Atmospheric Infrasound"
> *Physics Today*, vol. 53, no. 3, March 2000

1.1. NOISE IS AN UNWANTED SIGNAL

What is noise?

Noise is a nuisance. Noise is the hiss and pop of the static we hear when we listen to a radio station or a walkie-talkie or an old vinyl record album. It is the flickering snow that mars a cable TV image when someone bumps the cable or when the cable company scrambles the pay-per-view boxing match. It is the grainy streaks that flash across the screen when we watch an old print of a film. Noise is the background chatter of people in a restaurant when we ask the server to tell us about the dinner special. Noise is the Internet or telephone signal that interferes with our own. Noise is a red rose that grows in a cornfield. It is a signal that does not belong there.

Noise is a signal we don't like.

Noise has two parts. The first has to do with the head and the second with the heart.

The first part is the scientific or objective part: Noise is a signal. But then what is a signal? A mathematical answer is that a signal is what we describe with a variable such as *x*. That broad answer lets light or dollars or red blood cells count as signals because they can vary in time or space.

A physical answer deals with energy.

A signal is a source of energy such as an electrical pulse or a chemical pattern or the acoustical roar of an audience. It is structured energy. We scan the heavens looking for the light signals that bounce off dangerous asteroids or for the energy patterns that stream across empty space from stellar explosions or from alien civilizations. The military has tested "noise guns" that fire "sonic bullets" or tightly focused beams of sound energy such as a baby's crying played backwards at volumes so loud it rivals the sound of a jet taking off. Yet we see a signal as still more than energy.

A signal is anything that conveys information.[1]

All fields of science search for signals. Physicists look for particle signals in bubble chambers. Geologists look for earthquake signals in crust data. Botanists look for hormone signals in a pruned peach branch. Political scientists look for voter signals in polls and election results. Psychologists look for mating signals in barroom behavior and underarm sweat. Neuroscientists look for electrochemical signals in the vast synaptic webs of our brains.

The same holds for engineering but with a key difference: Engineers shape signals as well as search for them. Neural engineers design simple software brains that can detect buy-or-sell signals in the bond market or detect telltale signals from a buried land mine. Polymer engineers design molecular tubes that transmit light or electron signals. Civil engineers design buildings that bend or twist or even slide on rollers to dissipate the strain energy in earthquake signals. Electrical engineers (information scientists) design filters that let us watch someone's 3-D brain signals as he answers a test question or as he repeats the same answer in a foreign language. Communication engineers design complex codes of 0s and 1s that let us receive digital signals from deep-space probes and that let us see or hear the same signals on a compact disc or DVD even if we scratch the disc with a nail.

The second part of noise is the subjective part: It deals with values. It deals with how we draw the fuzzy line between good signals and bad signals. Noise signals are the bad signals. They are the unwanted signals that mask or corrupt our preferred signals. They not only interfere but they tend to interfere at random.

This signal badness differs from mere dislike of the signal content or the message that the signal conveys. Suppose you are in the large crowd in New York City's Times Square on New Year's Eve. People cheer and shout so loudly that you can barely hear your friend. Your friend's voice is the good signal while the other voice signals are bad signals. That does not mean that you like what your friend tells you. You may hate what your friend says ("I can't go") but that does not make it noise. Messages themselves are not signals even though signals can convey messages.

A more subtle issue involves politically correct speech signals. What you think are signals or messages you like may be noise signals that you would not like if you knew all the facts. That turns out to be the logical essence of so-called political correctness. Advisors to a king or boss or scientific journal editor may lie or distort their advice to enhance their reputation or to protect themselves from charges of racism or sexism or ageism or speciesism or any other point of view that society disdains or brands as politically incorrect.

Such subtle or gross lying can produce a "babbling equilibrium." This is a variation of the Nash equilibrium of game theory.[2] The advisor is so committed to pushing his agenda that his advice does not correlate with the facts. His advice is a random signal and the boss or decision maker comes to see this in time. So the boss ignores this advisor's signal as pure noise and just lets him babble. There is no information exchanged in such a babbling equilibrium and so society loses.

Yale economist Stephen Morris has used the mathematical theory of games to show that even good or unbiased advisors will lie this way with some probability if doing so will improve their reputation in the eyes of the decision maker.[3] The perverse result is that socially valuable signals about matters of fact and social policy can degrade into noise signals without society's knowing it.

So noise is a signal we don't like and signals consist of energy and convey information. Noise signals are bad signals or bad sources of energy. But for whom are they bad?

1.2. THE NOISE-SIGNAL DUALITY: ONE PERSON'S SIGNAL IS ANOTHER PERSON'S NOISE

Notions of badness vary from person to person. This theme runs throughout *Noise*: *One person's signal is another person's noise and vice versa.* We call this relative role reversal the *noise-signal duality*. A meteorologist or photographer may wait for days to see lightning strike a tall building. Both count the lightning discharge as a good signal. But a power engineer or nearby Internet surfer may count the same lightning discharge as a noise impulse. So the same energy counts as signal to one and as noise to the other.

The origin of the word "platinum" offers both a physical and a financial example of the noise-signal duality. The Spanish conquered South America as much for its rich deposits of gold and silver as they did to gain great land estates and to promote a Middle Eastern faith. Precious metals had washed down from the slowly eroding Andes over eons. Spanish prospectors and their Indian slaves often found small hard gray-white pellets at the bottom of their pans along with rare flakes and pellets of gold. The heavy pellets were a physical form of noise that interfered with the physical signals of the gold flakes and pellets.

The Spaniards called the troublesome pellets *platina* for "little sil-ver" from the Spanish *plata* for "silver."[4] Some prospectors even fol-lowed the Indian practice of throwing the *platina* pellets back into the river in the hope that the pebbles would somehow grow into gold. By the 1700s Spanish counterfeiters were using gold to coat platinum coins and ingots. The Spanish government went so far as to dump platinum coins in the ocean and ban the export of platinum from South America. But platinum pellets in a mine have long since passed from noise to signal. Today the price of platinum greatly ex-ceeds that of gold even though the earth's crust contains about ten times more platinum than gold.

A woman's heartbeat offers another example of the noise-signal duality. An electrocardiogram or ECG displays this signal as a se-quence of jagged pulses. The signal is a "time series" of data and is jagged because mild measurement noise corrupts the pure heartbeat signal. A major source of ECG noise is the 60 hertz noise from the power line. (A 60 hertz signal oscillates 60 times per second.) The woman can detect her own heartbeat signal if she takes her pulse or if she stands still in a quiet room and breathes softly.

Now suppose the woman is five months pregnant. She goes for a checkup and the physician lets her hear the heartbeat of her fetus by using an adaptive noise canceller. Her own heartbeat is now the major source of noise that the device must track and cancel. Her heartbeat can be as much as ten times stronger than the fetus's. And the fetus's heartbeat may follow the beat of a different drum-mer. It may beat faster or slower than the mother's heartbeat and its rate can fluctuate. The adaptive noise canceller uses each ECG measurement to build a mathematical model of the mother's heartbeat that changes as the heart data changes. The system sub-tracts this changing estimate from the ECG measurement. The subtraction cancels or at least suppresses the mother's actual heart-beat signal.[5]

Scientists and laymen may not agree on many things but most agree about noise. They may define or illustrate noise in their own way. But they share a common view of noise: They all want less of it. Few imagine a paradise that contains an active noise source.

It is not just that noise is a bad signal. And it is not just that we want to actively reduce noise. We all want cleaner images and more reliable data lines and more soundproof walls and windows. We want more: We want to eliminate noise. We want to wipe noise out of digital existence. We want to win the war on noise through total annihilation.

But we never will.

1.3. INFORMATION THEORY MADE A SCIENCE OUT OF THE WAR ON NOISE

The war on noise may be an undeclared war. It may not make headlines in the media. But it is surely one of the great technical conflicts of the information age.

Scientists and engineers have worked for over a century to rid our world of noise. They have found ways to measure and quantify it. They have designed systems to filter or subtract noise from a measurement. And they have proved that noise will always be with us.

The great leap forward came in 1948.

That was the year when Bell Laboratories scientist Claude Shannon published the landmark paper "A Mathematical Theory of Communication." This ingenious paper launched the modern field of information theory in almost one stroke.[6] It popularized the use of the terms "bits" or "binary digits" as units of information and showed how to measure the flow of information in terms of bits per second. The paper showed how a measure of entropy or disorder similar to that which physicists use to describe the thermodynamics of colliding atoms can also describe how many bits one can send or receive through a noisy channel. Much of the story of noise in the twentieth century is the story of information theory.

Shannon tried to solve one key problem in his paper: "The fundamental problem of communication is that of reproducing at one point either exactly or approximately a message selected at another point." One example is making sure that what you say on your end of the phone is roughly the same thing that your friend hears on her end. Shannon's seminal work made a science of information and spawned new fields of mathematics and engineering. The whole edifice rests on bits and so it makes sense to review the crucial concept of a bit of information.

The first question concerns origins: Just who came up with the name "bits"? Some scientists seem to think that Shannon himself coined this literal code word of the digital information age. But that is not so. Shannon's own 1948 paper gives credit for the term to the statistician John Wilder Tukey.

John Tukey (1915–2000) was one of the pioneers of the information age and deserves at least honorable mention here. He was at Bell Labs when Shannon worked out his information theory but later became a statistics professor at Princeton. Richard Nixon awarded him the National Medal of Science in 1973. Tukey was especially good at devising new words in engineering. He put forth the term "cepstrum" as a play on "spectrum" when he described a new way to model signals that one often encounters as a "chirp" in sonar or speech processing. Tukey coined the term "twiddle factor" as part of his joint discovery of what may be the most important computational algorithm in modern signal processing—the FFT or fast Fourier transform. The FFT acts on signals as a prism acts on white light. It breaks a time-varying speech signal such as how you pronounce "Tukey" into its component frequencies or spectral colors. Then we can shape or process the signal in the color domain rather than working with it in the original time domain. Removing noise amounts to removing some of the colors. The inverse FFT converts the processed colors back into a cleaned-up speech signal just as an inverted prism converts the rainbow of colors back into a beam of white light. Later research revealed that the eminent German mathematician Johann Carl Friedrich Gauss discovered the FFT algorithm as early as 1805.[7] He also discovered the bell curve of probability that bears his name and that we discuss at length in chapter 4.

The next question is more vital even if we seldom ask it: What *is* a bit?

A bit involves both probability and an experiment that decides a binary or yes-no question. Consider flipping a coin. One bit of information is what we learn from the flip of a fair coin. With an unfair or biased coin the odds are other than even because either heads or tails is more likely to appear after the flip. We learn less from flipping the biased coin because there is less surprise in the outcome *on average*. Shannon's bit-based concept of entropy is just the average information of the experiment. What we gain in information from the coin flip we lose in uncertainty or entropy.

But what if something rare happens? Doesn't that give us a lot of information? There is great surprise and thus a large gain in information when an improbable event occurs such as when David killed Goliath. But these rare events tend to wash out when we average over all possible events. That is why over time we learn more about boxing if we watch the world's two top boxers fight each other than if we watch the leading heavyweight boxer fight featherweights—even though the rare victory of a featherweight over the heavyweight would give us a bigger lone jolt of surprise.

Flipping a fair coin shares a key property with an equally matched sports contest: The outcome of each conveys one bit of information.

The nature of the contest itself does not matter in terms of information theory. All that matters is the *probability* of winning or of getting heads or of receiving a 1.

What does not matter to information theory is what often matters most to us—the meaning or content of the message. This reflects Shannon's famous statement in his 1948 paper that the "semantic aspects of communication are irrelevant to the engineering problem. The significant aspect is that the actual message is one *selected from a set* of possible messages" [emphasis in original]. There is some probability that the sender will send each binary message that we actually receive on the other end of the noisy communication channel. There may be a 70% chance that the sender will send a 1 and thus a 30% chance that the sender will send a 0. Information theory works with only these bits and probabilities.

Shannon saw that we could use bit *values* such as 0 and 1 to take a digital view of communication signals such as smoothly varying voltages or radio waves that are themselves analog in nature. How did he do this? He swapped the smoothly varying nature of physical signals for the smoothly varying probabilities that describe whether a binary signal is present or absent. The changing strength of a received electrical signal became the changing probability that the reader had read a 1. That swap turned the analog world of fact into the digital world of engineering.

That gave birth to digital communications.

Shannon's information theory was radical in the sense that it was fundamentally new. It summed up the past work on communication much as James Clerk Maxwell's famous four equations summed up work on electricity and magnetism in the nineteenth century. But Shannon's theory went far beyond a mere summation. Albert Einstein's theory of relativity was revolutionary because it overthrew Newton's theory of gravity. Shannon's theory was revolutionary because there was no real prior theory for it to overthrow. Even quantum mechanics and relativity relied on prior mathematical tools and concepts such as imaginary numbers and curved geometry. These fields did not introduce new math tools at their inception although their more recent versions have.

Shannon came up with new mathematical tools. Then he used these tools to answer questions that no one before had asked. Shannon achieved something as primal as Newton had when he (and independently Gottfried Leibniz in Germany) came up with the calculus. Newton put forth his theory of universal gravitation and then used the calculus and gravity to derive Kepler's three laws of planetary motion.

Shannon's new tools have since spread to all branches of science and engineering. Even physicists use his measure of entropy or disorder to estimate the information content of any object in the universe or of the universe itself. Throw the object into a gravitational black hole. Then the increase in the black hole's surface area measures the object's information content or "bit count." If you throw the whole universe into a black hole then you find that the universe itself has a bit count of about 10^{120} bits.[8] Note that here physicists are the "engineers" who apply the pure theory of the information "scientist" Shannon.

So if we had to pick a date and event where the information age began then a top contender would be 1948 and the publication of Shannon's paper. The modern information age arguably owes as much to Shannon's paper as it does to the rise of the computer and to Moore's "law" that the number of binary logic gates on computer chips has tended to double every two years or so. Indeed digital computers and chips are bit processors that merely apply the concepts in Shannon's paper and work within its bounds.

Noise lies at the heart of Shannon's theory because noise sets those bounds.

1.4. CHANNEL NOISE RANDOMLY FLIPS BITS

What Lord Acton said of power also applies to noise: It tends to corrupt. Noise corrupts a signal that flows through a channel. The signal carries a coded binary message of 1s and 0s that moves through the channel in one long list or in a group of smaller packets. The signal can be a spoken message on a telephone line. Or it may consist of the many binary packets that make up an e-mail message and that pass over different national or even international Internet routes on their way to the message receiver. The channel can be a phone or cable line or the wireless link through thin air from an earth-based radar dish to a satellite 22,300 miles above the earth in geosynchronous orbit.

The noise can come from the thermal jiggle of molecules in an electrical device or wire or from an electrical power surge or from clouds of solar wind that sweep through the ionosphere. Or the noise can be "crosstalk" from thousands of other signals passing through the same channel at the same time. The noise takes its toll on the message as it randomly turns some of the 1 bits into 0 bits and randomly turns some of the 0 bits into 1 bits: *Noise randomly flips bits.*

But is bit flipping the only type of digital noise?

Engineers often assume so. This can be a good working assumption because engineers try hard to minimize the so-called bit-error probability or the probability that channel noise will flip a single bit in a transmission. Scientists and engineers also try to maximize some type of signal-to-noise ratio such as the raw electrical power of the signal versus that of the contaminating noise. Increased signal power tends to reduce this probability of error and thus it tends to improve how well our eyes or ears or machine sensors can detect signals in the presence of noise. Redundant coding schemes can also reduce the bit-error probability. But redundant codes transmit less information just as we do when we speak on a noisy phone line and repeat each word so that the receiver will be sure to hear what we say. All channel noise undermines to some extent the reliable transmission of bits—and thus noise still tends to flip bits.

But noise can be subtler in a digital system. Noise can disturb the *timing* of when a bit value arrives at a receiver as well as randomly flipping that bit value. Every teenager has figured out that he can send messages to friends without paying for them. John can work out a simple code that lets his friend's phone ring a certain number of times. A long and dumb code would let the phone ring five times for the fifth letter in the alphabet and twenty times for the twentieth letter and so on.

A little more thought shows that John need let the phone ring only once if he works out a code based on the time between successive calls. But even here noise strikes if John makes calls to his friend too close together in time. The rings may confuse his friend if John does not carefully watch the second hand on his watch to get the timing right. John's timing errors or sloppiness produce timing noise. This noise may be random but it does not flip bits.

Timing noise would also arise if John tried to be clever and mailed empty envelopes to his friend every second or third day. Erratic postal service would compound the timing noise. And timing noise arises in the billions of bit packets that race through Internet channels at or near the speed of light.

Indeed such timing noise afflicts every neuron in our brain and body. Each neuron acts as a type of binary switch. Electrical charge builds up in the neuron just as water builds up behind a dam. The electrical dam bursts and the neuron emits an electrical spike or "action potential." The spike or its absence defines a flesh-based digital signal of a 1 or 0. The neuron rests a moment and then starts to build up energy again as thousands or even millions of other spikes flow into it from other neurons in the neural network.

The result is that each neuron emits a binary *spike train*. The spikes have the same height or energy but their timing or position in the spike train differs. Neurons convey information to one another through the timing of their spikes. The average neuron in the brain broadcasts its spike train to about 10,000 other brain neurons. That helps explain why brains have so much "gray matter": It takes a lot of wet wires or biochannels for a hundred billion spiking neurons to talk to one another. Each spike train involves its own timing noise and the massive feedback structure of neural circuits greatly compounds this noise. Neurons may use redundant coding with multiple spikes to combat this structural source of noise. And yet a single neural spike conveys information just as the eye can respond to a single photon of light.[9] The great research goal of computational neuroscience is to make information sense of a brain's billions of neural spike trains.

Most of twentieth-century information theory ignored the timing issue and so it ignored timing noise. An important exception came in 1996 when Berkeley electrical engineer Venkat Anantharam and Princeton engineer Sergio Verdu published a paper called "Bits through Queues." Their paper showed that the timing of digital messages could increase the amount of information that flows through service channels such as Internet pathways or John's phone line or even supermarket checkout lines. It won the prestigious Information Theory Society Best Paper Award in 1998 and opened a new line of noise-based research in queuing theory.[10]

Shannon ignored timing noise and focused instead on how many bits one can *reliably* send through a channel. He proved that noise limits how much information can pass through a channel if we want to achieve a given level of accuracy in a given length of time. Channel capacity constrains reliable transmission.

This is the so-called channel capacity theorem. No person or computer or space alien can reliably send more bits over a channel at a rate higher than the channel's capacity. This mathematical fact seems to be a law of nature and some version of it appears to hold at the quantum level. Noise damns us to limited capacity. It constrains how much information can reliably pass over the Internet or through the skies or down the spike-train pathways we call our nerves. Stanford information theorist Thomas Cover sums up the theorem this way: "The Channel Capacity Theorem is the central and most famous success of information theory."[11]

The channel capacity theorem also has a positive side. It says that in theory all is fine if you don't exceed the capacity. If you work hard enough or smart enough then you can in theory find ways to code signals so that you can send them in effect without error—provided again that your bit rate does not exceed the channel capacity. Shannon showed how some codes picked at random could achieve such ideal near-zero error rates. (This means there need not be an infinite regress of schemes to correct errors that might arise because each such scheme could produce its own errors.) So far no one has found a practical way to construct these ideal near-zero error codes although recent heuristic codes called "turbo codes" do come close to the mark. The identity of these ideal or optimal codes remains the great open question of information theory—even if turbo codes or other suboptimal codes suffice in practice.

An understanding of Shannon's achievement requires a closer look at a channel's information capacity.

1.5. NOISE LIMITS CHANNEL CAPACITY

A digital communication channel is a bit pipe in direct analogy to a water pipe. The channel capacity corresponds to the size of the bit pipe. Only so many bits can reliably flow through a channel per second just as only so much water can pass through a water hose per second without bursting the hose. But we can do better than describe a channel by mere analogy.

We can define the channel capacity in terms of bit information and probability descriptions: Channel capacity maximizes the *mutual* information between sender and receiver. It maximizes this term over all possible probability descriptions of the sender's bit-based experiments or transmissions. Real and potential noise create these probability descriptions. A set of conditional probabilities describes the noisy channel itself. These probabilities describe the chance that you will receive one type of message (such as your friend saying "yes" over the noisy phone line) if the sender sent some other message (your friend in fact said "no").

Then what is mutual information? Shannon defined it as what we learn from getting an answer or receiving a message. Mutual information measures an uncertainty gap or entropy gap and thus an information gain. It measures how much on average the received symbol 0 or 1 reduces the receiver's original uncertainty that the sender sent a 0 or 1 through the noisy channel.

Suppose John flips a coin and then sends you either the bit value 0 or 1 through a noisy channel. What is the probability that John sent a 0 if you receive a 0? What is the probability that John sent a 1 if you receive a 0? You don't know. It depends on the probability description of what John sent. One such description is this: "There is a 60% chance that John sent a 0 and a 40% chance that he sent a 1." Your uncertainty about what John sent is your entropy.

Consider a minimal noisy channel or a so-called binary symmetric channel:

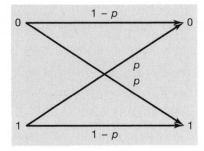

Figure 1.1: The binary symmetric channel. If there is a probability p that the noisy channel will flip a 0 bit value to a 1 bit value then there is probability $1-p$ that the channel will not flip the 0 bit value. The same probabilities describe whether the noisy channel flips the 1 bit value to a 0 bit value.

The figure displays how a noisy channel can flip either bit value that John sends through it. Suppose there is a 60% chance that the channel will flip the 0 bit value. Then there is a 40% chance that it will not. (The same would be true of flipping the 1 bit value because the channel is symmetrical.) This simple figure captures the logical and random essence of Shannon's information theory. The signals are digital while the probabilities are analog or continuous. The great superstructure of information theory thus rests on probability theory.

Now suppose you receive a 0 from John. You are still not sure which bit value John sent when you receive the 0 but you have more information than if you know only that John sent something. The channel may have been so noisy that the bit flipped a hundred times before you got it. You have no way to know. The presence of noise forces us to give up knowing such things with certainty. You just know that John sent something and that you received the 0 bit value. You are in a state of so-called *conditional* entropy because your uncertainty about what John sent depends on what you received. You would be in the same state of conditional uncertainty if you had received a 1. But it is no more uncertainty than before you received anything at all.

The key point is that getting the message tends to reduce your uncertainty. The received message is experimental data. Receiving it creates a gap between your present conditional uncertainty and your prior unconditional uncertainty. That gap is exactly your gain in information. It turns out that John would on average experience the same gap if you sent him one of the bit values 0 or 1 through the same noisy channel. That is the mutual or symmetric part of mutual information. It is symmetrical between the source and receiver.

This reflects a mathematical fact: Mutual information is the receiver's entropy minus the conditional entropy of what the receiver receives—given what message the sender sends through the noisy channel.[12] Conditioning or getting data can only reduce uncertainty and so this gap is always positive or zero. It can never be negative. You can only learn from further experience. Information theorists capture this theorem in a slogan: *Conditioning reduces entropy*. The channel capacity itself is the largest gap given all possible probability descriptions of what John sent. It is the most information that on average you could ever get out of the noisy channel.

Each bit also comes at a cost. The mere presence of "thermal" or molecular noise implies that there will never be a bit-based free lunch: It takes a tiny but minimal amount of energy to transmit a single bit of information. It can take even more energy in practice. (The tiny energy minimum is ln2 kT joule or 0.693 kT joule where T is the noise temperature in kelvins and k is Boltzmann's constant of 1.38×10^{-23} joules per degree. That is far less than a trillionth of a trillionth of a kilowatt-hour.) So noise limits the information in anyone's information age. Chapter 4 discusses thermal noise and other types of noise.

This limit comes from Shannon's channel capacity theorem: Channel capacity depends on the strength of the corrupting noise as well as on the transmission bandwidth and the strength of the signal. *Noise limits channel capacity.*

Shannon showed that even infinite bandwidth does not overcome the problem of noise because noise grows in direct proportion to bandwidth. Some media-age gurus have overlooked this bitter fact of noise: Infinite bandwidth does *not* mean infinite capacity. We would still need infinite signal broadcasting power to transmit an infinite amount of information over a communication channel. Infinite bandwidth would buy us only a *linear* gain in information capacity that is directly proportional to our finite signal power. Researchers Partha Mitra and Jason Stark of Bell Labs have shown that noise takes an even higher toll on fiber optics because the channel is no longer linear but nonlinear. The gain in capacity soon reaches a maximum and then *declines* if we increase the signal power any further.[13]

So we remain creatures of limited capacity even if we somehow had infinite bandwidth. We can thank noise for that. And we can thank noise for limiting the bit rate of our phone lines to about 22,000 bits per second. This in turn limits our increasingly Internet-based society.

Noise seems to cause only trouble. Chapter 3 discusses how noise harms our hearing and overall health. And Shannon's results alone say that we can never win the war on noise. We might find optimal ways to wage the war in special cases. Yet even here we struggle just to achieve a noise-based limit. So it makes sense to explore defeat: Does noise win the war in the end?

It does in at least one important sense: cosmic death.

Physicists have long predicted that the universe will end in a cold "heat death" or more properly a "cold death." This is the final triumph of the inexorable second law of thermodynamics which says that the physical entropy or *average* disorder of a closed system only increases in time. The ordered molecules of matter and of life itself are but temporary patterns on the way to random disintegration.

The building blocks can come together for a while but sooner or later they will come apart and end up in random disarray. And heat itself is but one form of noise.

The evidence so far is that the universe will expand forever and so end in a random cold death.[14] There is not enough matter to make the universe contract someday under its own gravitational self-attraction. So star formation will end. And stars themselves will burn out after billions or trillions of years—and then things will grow colder and darker still. The average organization of molecule swarms will grow less ordered and more random. There will be less structured energy even though the total amount of energy stays constant. So there will be fewer signals or at least fewer "good" signals. Physicists describe this likely outcome in melancholy terms:

> The universe as we know it, with shining stars, galaxies and clusters, appears to have been a brief interlude. As acceleration takes hold over the next tens of billions of years, the matter and energy in the universe will become more and more diluted and space will stretch too rapidly to enable new structures to form. Living things will find the cosmos increasingly hostile.[15]

Living things will also find the universe noisier—quieter and colder yet noisier.

1.6. NOISE CAN SOMETIMES HELP

But is all noise bad?

This question does not restate the noise-signal duality that one person's noise is another person's signal and vice versa. That view sees the badness of a noise signal as relative. This is an absolute question.

Consider simple random interference with a signal. Is it always bad? Are there cases where we would *deliberately* seek such interference?

The surprising answer is yes: *Noise can sometimes help as well as hurt.* Much of this book attempts to unpack that last statement. The chapters that follow will explore both the negative and the positive aspects of noise. The last chapter focuses just on noise benefits.

For now consider this summary in *Nature* on the constructive use of noise from Rutgers biochemical engineers Troy Shinbrot and Fernando Muzzio:

> The use of noise for productive purposes has a long and colorful history. For over a century it has been reported by sailors that disordered raindrops falling on the ocean will calm rough seas. By comparison, noise in the form of heat has for many decades been used in the electronics industry to instigate charge release from surfaces and so amplify signals (in vacuum tubes), deposit ions (in thin-film fabrication), or generate displays (in cathode ray tubes). Noise in the form of tiny disturbances provokes instabilities that lead to regular oscillations in systems ranging from musical instruments and humming power lines to road corrugation and catastrophic structural collapse. Noise is also important in increasing signal processing effectiveness in applications as diverse as neuroprocessing and paleoclimatology. And the addition of noise to pattern-forming systems can actually enhance the pattern-formation mechanism in several problems. In other applications, aspects of noise can improve the regularity of longterm biological rhythms.[16]

Next consider this left-to-right sequence of four noisy images of the *Playboy* Playmate Lena:

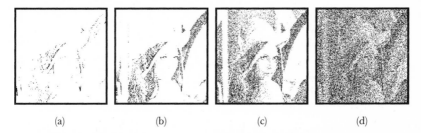

(a)   (b)   (c)   (d)

Figure 1.2: "Stochastic resonance" noise benefit. Image contrast improves and then degrades as the level of random pixel noise increases from no noise in the leftmost image to maximal pixel noise in the rightmost image.

"Lena" has long since become the standard benchmark in image processing. My former graduate student Sanya Mitaim and I used this example to show engineers and scientists that random pixel noise could in fact improve our subjective perception of an image.[17] We started with the standard image and reduced almost all of its contrast. That produced the image in frame (a). Then we randomly added uniform or equiprobable noise to each pixel or picture element. That produced the improved image in frame (b). More intense pixel noise produced the even more improved image in frame (c). Still more intense pixel noise degraded the image as in frame (d). So the image quality *increased* when we added a small amount of noise but then decreased when we added too much. Our paper was the first to show how neural learning schemes could find the optimal amount of noise for many systems—and that optimum is seldom zero noise.

The general name for this type of noise effect is *stochastic resonance*.[18] European researchers proposed this awkward term in the 1980s to describe how physical noise from solar variations to ocean fluctuations might explain the 100,000-year periodicity of the earth's ice ages. This remains an active and controversial hypothesis.[19]

Stochastic resonance (SR) soon came to refer to any case where a small amount of noise can improve how a system detects faint signals.

Physicists and biologists have since founds hundreds of examples of SR in nature. Sharks use noise to catch prey while crickets use air noise to detect killer wasps. The largemouth bass is the most popular freshwater game fish in the United States. Its favorite food is the humble crayfish. Biophysicist Frank Moss at the University of Missouri in St. Louis showed that the crayfish's tailfin sensors use the noisy chaotic froth of water to help them detect the periodic back-and-forth sine-wave-like wiggle of largemouth bass.

Moss went on to show that the long-snouted river paddlefish exploits faint fields of electrical noise when it feeds on tiny zooplankton in murky rivers and lakes. The paddlefish has survived and evolved for about 300 million years—which makes it the oldest animal in North America. It is a close cousin of the sturgeon and poachers often swap its roe for sturgeon roe (caviar). A swarm of tiny plankton jointly emit an electrical field that acts as noise. The paddlefish's long snout acts as an antenna because it contains tens of thousands of tiny passive electrosensors similar to those found on the "pores" of sharks and rays. Sharks can use these so-called ampullae of Lorenzini to detect the faint electrical fields from sea creatures hiding beneath the sand on the ocean floor.

Moss studied how electrical noise affected paddlefish feeding in a tank. He showed that too much plankton noise confuses the paddlefish and so it catches fewer plankton. Too little plankton noise also leads the paddlefish to catch fewer plankton. The optimal amount of electrical noise lies between these two extremes and lets the paddlefish catch the maximal amount of plankton.

Neural tissue may also exploit noise.

The billions of neural spike trains in our brains produce a ceaseless atonal symphony of noise as if billions of musicians played their instruments independently and as if each of the musicians broadcast his music to ten thousand other musicians. Yet model neurons seem to use noise to maximize their ratios of signal to noise or their over-all Shannon mutual information.[20] Our minds may somehow have emerged from this neural cacophony of billions of noisy neurons. And our minds likely do so not just despite all this noise but with the aid of it.

So it is not just that we will never win the war on noise because noise is in the physical nature of things. Careful analysis shows that in many cases we should not even be fighting it.

CHAPTER 2

As we continue to attack offenses such as prostitution and drug dealing to improve New Yorkers' quality of life, we must also target other chronic and disruptive problems like noise.

    −New York City mayor Michael Bloomberg
    2 October 2002

Just as in the case of the barking dog and the noisy party, the annoyance is usually too petty for formal and expensive proceedings in equity and ought, in all reason, to be settled by mutual concessions and neighborly good will.

    −William H. Lloyd
    "Noise as a Nuisance"
    University of Pennsylvania Law Review, vol. 82, April 1934

Section 112.01. Radios, Television Sets, and Similar Devices.

(a) It shall be unlawful for any person within any zone of the City to use or operate any radio, musical instrument, phonograph, television receiver, or other machine or device for the producing, reproducing or amplification of the human voice, music, or any other sound, in such a manner, as to disturb the peace, quiet, and comfort of neighbor occupants or any reasonable person residing or working in the area.

(b) Any noise level caused by such use or operation which is audible to the human ear at a distance in excess of 150 feet from the property line of the noise source, within any residential zone of the City or within 500 feet thereof, shall be a violation of the provisions of this section.

(c) Any noise level caused by such use or operation which exceeds the ambient noise level on the premises of any other occupied property, or if a condominium, apartment house, duplex, or attached business, within any adjoining unit, by more than five (5) decibels shall be a violation of the provisions of this section.

–Noise regulations

Los Angeles, California: Chapter XI

Ordinance Number 156,363

Noise Free America has the following agenda: Each State of the United States is called upon to declare noise a dangerous form of pollution, a serious threat to health and safety, and a widespread problem subject to State jurisdiction. All States should adopt a comprehensive Noise Code which would form the general framework for county and local ordinances within the State. Counties and localities would be required to adopt the elements of the Noise Code and would be required to submit any allowable local modifications to the State for approval. No element of the Noise Code could be disallowed by the local jurisdictions. However, any local modification of the Noise Code or separate regulation construed to be more stringent than the State Noise Code would be allowed.

–Noise Free America Web site

The liberty of the individual must be thus far limited: He must not make himself a nuisance to other people.

–John Stuart Mill

On Liberty

Nuisance doctrine is an ideal legal mechanism for forcing spammers to in-ternalize the costs they impose on innocent third parties.
—Adam Mossoff
"Spam—Oy, It's Such a Nuisance!"
Berkeley Technology Law Journal, vol. 19, Spring 2004

What do you do if your neighbor plays her music too loudly?

One approach is to knock on her door and ask her to lower the volume. That is socially awkward and may lead her to slam the door in your face and play her music even more loudly.

Another approach is to be an equally bad neighbor and play your music loudly in response. But that can interfere with your concentration or even damage your inner ear—as we discuss in the next chapter. Noise has permanently damaged the hearing of about 10 million Americans and produced varying degrees of stress and annoyance in many millions of others.

An extreme approach is to be a very bad neighbor and sue her in civil court for the tort of private nuisance. Energetic noise signals can thus invoke law and medicine as they pass from mere annoyance to actionable interference or biophysical hazard or even environmental threat. And thus noise has emerged as a new and important frontier for social activism.

2.1. NOISE IS A PRIVATE NUISANCE IF IT SUBSTANTIALLY AND UNREASONABLY INTERFERES WITH SOMEONE'S USE AND ENJOYMENT OF LAND

Noise is a nuisance because it is a bad signal that interferes with a good signal or that interferes with the absence of certain sound signals that we call peace and quiet. But unwanted sound or auditory noise can constitute a literal nuisance in the sense of the common law of torts. This requires a detour into the intricate legal theory of nuisance. Torts grandmaster William Prosser warned of this intricacy: "There is perhaps no more impenetrable jungle in the entire law than that which surrounds the word 'nuisance.' It has meant all things to all people and has been applied indiscriminately to everything from an alarming advertisement to a cockroach baked in a pie."[1]

Sound noise often rises to the level of nuisance. Sometimes this results from living too close to a commercial enterprise or to a suburban neighbor who raises chickens or who uses his garage as a practice room for his new rock band. The sound energy of a source falls off with the inverse square of the distance from the source. So even a few meters can make a difference if you live near a dairy farm or a busy parking lot or a go-cart racetrack. The courts have even entertained noise nuisance suits for the loud ringing that results when someone makes too many unwanted phone calls. One court found a nuisance because "we cannot say that calls late at night and from three to seven calls at the dinner hour for several months would not distress a person of ordinary habits and sensibilities."[2]

The economic structure of a nuisance is a local market failure. Playing a boom box loudly at a crowded beach produces a benefit for the user but imposes unwanted costs or so-called negative externalities on third parties. This results in a local dispute that may have no optimal resolution. The user argues for his positive liberty of action while the third party argues for her negative liberty or right to be let alone.

The Nobel Prize–winning Coase theorem of economics says that the two parties can always work out their differences if the transactions costs of doing so are negligible and if it is crystal clear who owns what. Ronald Coase first published his theorem in 1960 in the *Journal of Law and Economics*. He based his analysis in part on an old 1879 lawsuit called *Sturges v. Bridgman* that involved the noise nuisance that a doctor suffered from a next-door candy maker. Economists call this ideal state a Pareto efficient outcome or a Pareto equilibrium. A Pareto equilibrium is such that no state or person or God could make any one person better off without making at least one person worse off. One ideal Coase result may be that the noise polluter turns down his boom box or pays the third party for the privilege of blasting his music. Another Coase outcome may be a Pareto equilibrium that the parties reach through side payments as when the third party pays the blaster to stop blasting.

Both conditions of the Coase theorem often fail in practice. The first condition of negligible transactions costs can fail when you face asking a hostile stranger to turn down and keep down her boom box. Many people find that too awkward or simply will not risk a hostile response. The second condition of clear (binary) property rights can fail because there are no well-defined property rights for the sound waves that pass through the air and through our bodies. The boundaries between mine and thine are too gray or fuzzy. The optimal outcome of the Coase theorem may hold only approximately if either of the theorem's two conditions holds only approximately.[3] The result is that the heavy hand of the law tries to balance the competing interests and to match costs to benefits. The common-law tort of private nuisance has evolved from hundreds of years of such utilitarian compromise.

A nuisance resembles a trespass. Both are torts or "civil wrongs" that affect someone else's land or property without the owner's permission. A trespass requires a physical invasion from a person or animal or even from a rock that someone throws onto the land. A nuisance occurs only if the invading object is less tangible than a rock or dirt clod. The invading substance can be dust or smoke or sound vibrations—and so it can be noise. It can be someone else's energy that you don't like. This prohibition on imposing unwanted signals on a neighbor reflects the old common-law maxim that underlies the tort of nuisance: *Sic utere tuo ut alienum non laedas*—use your own so as not to hurt others.

Trespass and nuisance also differ in where the actors act. You create a nuisance for your neighbor only if you do something objectionable on your own land or property. Blasting your stereo is a common example. That activity affects your neighbor's land or property or his use of it. But you trespass when you go onto his land. So you stay at home to be a nuisance but you go next door to trespass.

Nuisance is a form of indirect trespass.

Both nuisance and trespass have common roots in the old common law of England that extends back at least to the thirteenth century and focuses on land as power. Vestiges of that focus on land remain as the modern civil and criminal laws that impose strict liability for trespass—for the intentional and unprivileged invasion of the property of a lawful possessor.

Private nuisance grew out of the broader cause of action for *indirect* trespass called "trespass on the case." Prosser states that this distinction between direct and indirect trespass still persists: "The distinction which is now accepted is that trespass is an invasion of the plaintiff's interest in the exclusive possession of land while nuisance is an interference with his use and enjoyment of it. The difference is that between walking across his lawn and establishing a bawdy house next door, between felling a tree across his boundary line and keeping him awake at night with the noise of a rolling mill."[4] The indirect nature of a nuisance also explains why a plaintiff can sometimes prevail on both a nuisance claim and a trespass claim for the same underlying event such as flooding land or rock blasting.

Economists use the Coase theorem to distinguish nuisance from trespass. They often focus on the question of whether transactions costs are negligible rather than on whether the second prong of the theorem holds—whether the property rights involved are well-defined or binary.

A typical result is that trespass laws are efficient in terms of maximizing the whole social economic pie if transactions costs are negligible. Nuisance laws are efficient if the transactions costs are not negligible: "when transactions costs are low, the law of trespass should govern, and when transactions costs are high, the law of nuisance should govern."[5] The transactions costs involved in asking a neighbor to get out of your yard are low compared with getting a nearby factory to reduce its noise level. The higher transactions costs involved in a legal nuisance also explain why judges use more discretion in nuisance cases than in trespass cases. Some law-and-economics scholars have argued that a little bit of fuzz in property rights helps smooth out such nuisance disputes between neighbors. A common example is a working policy of live and let live where "individuals have to put up with a certain amount of noise and interference from their neighbors on condition that the neighbors reciprocate in kind."[6]

Nuisance also differs from trespass in damages or remedies.

A plaintiff can always get damages if he proves trespass. This also hails from the feudal days of England when land was king and when the king owned a good deal of the land. The damages may be the mere nominal award of one dollar. Even that award can declare who owns the land and so clear a clouded title or it can give some token emotional satisfaction. The court can award money damages that compensate for actual damage as when a child throws a rock through a glass window and then the rock proceeds to break a rare Ming vase. The court can award punitive damages in extreme cases of willful trespass. The court (sitting "in equity") can issue an injunction that forbids any further acts of trespass. The court can also force a defendant to pay restitution for any unjust benefits that the defendant received from the trespass as when a trespassing tractor routinely drives through a long strip of land rather than around it.

Damages are not automatic for nuisance even if the plaintiff proves that the noise or other interference amounts to a nuisance. The court instead *balances* the costs and benefits of the nuisance. So the court may let a noise polluter keep on blasting if the economic gain from local jobs or tax revenues outweighs the cost to nearby neighbors or pedestrians. And everyone has a liberty interest in pursuing peaceful activities that produce some amount of third-party noise.

Judges have to draw lines through such activities when they balance the relevant costs and benefits of a nuisance suit. The hazard of judicial balancing is one more reason why it may be better just to politely ask your neighbor to turn down the volume rather than ask a court to award money damages or issue an injunction. A court may grant no remedy even if you prove your case—and that outcome may only encourage the noisemaker to make more noise.

The formal legal definition of a private nuisance is a substantial and unreasonable interference with someone else's use and enjoyment of their land or other real property such as a home or apartment complex. The term "substantial" is just legal talk for big or large or a high level. The term in nuisance cases often means that the interference is big enough to reduce the property's market or rental value. Prosser states that "a good working rule would be that the annoyance cannot amount to unreasonable interference until it results in a depreciation in the market or rental value of the land."[7]

The term "unreasonable" in law tends to mean that average costs outweigh average benefits. The law often imagines a disembodied "reasonable person" to judge "objectively" whether a defendant's actions were reasonable under the circumstances in the sense that the person's expected benefits exceed his expected costs. This is the heart of the law of negligence or whether unreasonable conduct causes foreseeable harm. A car driver acts unreasonably if he swerves out of his lane simply to get a better view of an accident and runs over an accident victim who lies bleeding to death in the middle of the road. The driver acts reasonably in running over the accident victim if he swerves out of his lane to avoid a head-on collision with an out-of-control oncoming truck. The law assumes that we each value our own life more than we value the life of a stranger.

The doctrine of private nuisance applies the term "unreasonable" to the property owner and not to the defendant noisemaker. The noisemaker's conduct may well be reasonable in the sense that her local benefits from the loud music exceed her costs in terms of her personal balance of pleasure to electricity costs or inner-ear damage. The cost-benefit calculus applies instead to the property owner. And again the result is often that the harm must affect the property or rental value. The noisemaker can still always defend by claiming that the plaintiff assumed the risk of living near the noise or at least that the defendant impliedly consented to the noise. This holds even if the plaintiff can show that the noise satisfies all the elements of the tort of private nuisance.

The interference in a nuisance action is noise if it is an unwanted signal and hence if the interference is unwanted energy. It is again often sound energy as when the next-door neighbor practices her drums or tunes her Jet Ski or holds raucous Saturday-night parties. Or the interference can be the chemical energy of unwanted livestock molecules that strike the nose. An example of both types of unwanted energy is the private nuisance of a pigpen next door where the gregarious hogs grunt and excrete in ceaseless variety. Much nuisance law has dealt with piggeries (as in the 1917 case of *Clark v. Wambold*[8]) in the early twentieth century because of the growth in urban land areas and the resulting border clashes between city and rural folk. Piggeries can sometimes rise to the level of a public nuisance that we discuss below. Zoning laws soon codified nuisance principles into many of the municipal ordinances that govern modern life in the city.

Zoning laws and ordinances must change to keep up with

changing technology and the resulting new unwanted signals. Modern man consumes more electricity than ever and uses part of it to blast more signals through the air to third parties than ever—and so creates more private market failures than ever. Noise ordinances make it easier for citizens to get noise relief without resorting to the expense and risk of a private court action even if they offer little relief for the occasional nighttime car alarm. The epigraph above shows part of a noise zoning law for Los Angeles. The Noise Pollution Clearinghouse's Web site lists other noise ordinances from major cities in the United States. Concerned citizens should at least read them.

2.2. NOISE IS A PUBLIC NUISANCE IF IT SUBSTANTIALLY INTERFERES WITH A RIGHT COMMON TO THE PUBLIC

A neighbor's loud stereo can create a private noise nuisance if he plays the stereo frequently enough. But the local airport or strip bar can rise to the level of a *public* nuisance if it substantially interferes with the public's use and enjoyment of their land.

Public nuisance differs from private nuisance in the scope of the nuisance and in who can oppose it in court. It still involves a market failure: Someone benefits from an activity that imposes unwanted signals on third parties. But the scope in this case must cover several members of the public. This involves a fuzzy border with large-scale private nuisances but courts still make the distinction. A public nuisance further requires in general that only the attorney general can bring a legal action for public nuisance in a court of law or equity. The idea is that the attorney general litigates on behalf of the public. The same reasoning permits an attorney general to enforce the terms of a charitable trust.

A lone member of the public cannot sue over public nuisance

unless he can show that he suffers some special harm that the other members do not suffer. Proximity to the nuisance may sometimes satisfy this condition if the new smokestack factory is next door to his house and to no one else's. Even then the person must not have "come to the nuisance" or have moved to the house after the factory had polluted. And even then the court may balance the factors against him if the judge decides that the factory does more public good than harm. The result can be an innocent person's suffering harm for an alleged commercial good of the public. Such apparent injustice can easily spawn "social activism" and indeed has fueled some of the environmental movement as well as local community protests over new land development. A leader in this movement is the Noise Pollution Clearinghouse based in Vermont. The related League for the Hard of Hearing based in New York sponsors the International Noise Awareness Day on April 24 of each year.

A crematorium gives an eerie example of a public nuisance. The local crematorium burns corpses and then grinds the remains into powder. The burning process can take up to three hours and can produce a variety of unwanted signals that in some cases can rise to the level of a public nuisance.

The crematorium may not produce unwanted sounds or even unwanted sights because of its engineering and architectural design. But it likely produces unwanted chemical signals and some of these noise signals smell and some do not. Burning dead bodies gives off clouds of sulfur dioxide and nitrogen oxides as well as carbon monoxide and carbon dioxide that can irritate the lungs or eyes. It also turns some teeth fillings into dangerous vaporized mercury and can produce trace elements of the toxic heavy metals cadmium and lead. The 1990 amendments to the Clean Air Act require that the Environmental Protection Agency monitor the chemical pollutants of crematoria but this can be difficult to do in practice. Each state has laws that govern the licensing and inspection of crematoria. These laws may offer little comfort to local residents. Still an application to build a new crematorium can liven up a meeting of the local zoning board. And few real estate ads boast of a home's proximity to a crematorium.

The cremation noise problem will only increase in time as the population grows and as the cremation rate rises. The funeral industry largely opposed cremations until the 1960s. Most revenue and profit in a burial-based funeral stems from the wood or metal casket that houses the formaldehyde-infused corpse. But most cremations use only a cardboard or cheap lumber box to transport the deceased to the crematorium's gas flames where the local temperature can reach 1800 degrees Fahrenheit. This is so hot that the carbon-based corpse ignites and combusts. Only 10% of Americans were cremated in 1982. That figure rose to a full 25% in 2002 when there were 1,825 crematoria in the United States—and almost 10% of these or 177 crematoria were in California. Some experts estimate that the cremation rate will rise as high as 45% in 2025[9] as more aging baby boomers choose cremation over burial and over the far more rare cryonic suspension in liquid nitrogen.[10]

A more litigated public nuisance involves low-flying aircraft near airports. The same powerful engines that lift and propel metal carriers in the sky can emit highly energetic noise signals that propagate for miles. Flight paths can amount to an airport's taking an aerial easement in the airspace over a landowner's land. A city or municipality need not formally take the easement under its police powers of eminent domain. But the overflights may have the same effect of such a taking and thus constitute an inverse condemnation of the airspace.

The Supreme Court opened the door to inverse-condemnation noise challenges in the 1946 case of *United States v. Causby*. So many military airplanes flew over a chicken farm that it could not function. The Court ruled that the overflights amounted to a Fifth Amendment government taking and so required the government to compensate the farm owner. The Court extended this reasoning in the 1962 case of *Griggs v. Allegheny County* when it found that the county airport had taken an air easement for takeoff and landing over a homeowner's adjacent land. Pennsylvania State Supreme Court chief justice Bell summarized the noise effects on the homeowner when that lower court heard the case:

> During these flights it was often impossible for people in the house to converse or to talk on the telephone. The plaintiff and the members of his household were frequently unable to sleep even with ear plugs and sleeping pills. They would frequently be awakened by the flight and the noise of the planes. The windows of their home would frequently rattle and at times plaster fell down from the walls and ceilings. Their health was affected and impaired and they sometimes were compelled to sleep elsewhere.[11]

Landowners near airports often bring inverse-condemnation lawsuits for noisy overflights but with mixed success. Courts will not find a government taking if the noise behavior still allows the landowner what the Supreme Court calls "economically viable use of his land" and if the noisy behavior "substantially advances legitimate state interests" such as giving the public a convenient airport hub. Sometimes the government will buy local homes or move the residents rather than move the noisy airport. The Louisville International Airport in Kentucky set a precedent in 1999 when it dissolved the airport suburb of Minor Lane Heights and moved the whole suburb to a new area eleven miles away.[12]

Local landowners also lose their noise battles under the doctrine of a *prescriptive* aeronautical easement. Prescriptive refers to the squatter's-rights doctrine of adverse possession: A stranger starts using land without the owner's permission and then the stranger eventually comes to own the land if the original owner does not eject him. The rationale for this ancient doctrine of quasi-theft is that it keeps land from lying fallow and so promotes a higher economic use of the land. It also gives the owner an incentive to keep his title clear in our complex recording system. Frustrated Malibu beachfront owners confront this doctrine in the form of a prescriptive public easement when the unwashed masses continue to walk across their land to get to the cold green Pacific Ocean teeming with pollutant-nourished microbes.

The aerial version of the doctrine of prescriptive easement applies to the tens of thousands of us who live in the westerly approach path of Los Angeles International Airport. Rumbling jumbo jets fly overhead in a staggered "string of pearls" as they use this public easement that cuts through both public and private airspace. Our only remedy is to grin and bear it.

2.3. E-MAIL SPAM COUNTS AS A CYBER-NOISE NUISANCE

Spam is junk e-mail. It is unwanted e-mail from commercial advertisers and sometimes from nonprofit sources such as those who promote charitable events or political causes or professional conferences.

The amount of spam has grown exponentially over time. And the nature of many spam ads has grown more lurid with ads that promote a wide range of sexual goods and services as well as financial opportunities and pharmaceutical products. A sophisticated spammer can send millions of spam e-mails each day. So billions of spam messages traverse cyberspace each day. The amount of blocked or filtered spam has also grown exponentially but so far no anti-spam system has been able to block all unwanted e-mail and accept only wanted e-mail. Spam can come in too many forms for that and spam advertisers can be devilishly clever in how they structure and distribute their e-mail messages.

Spam counts as noise in cyberspace because e-mail messages are digital signals and because most computer users don't want them. It makes sense to call spam *cyber-noise* because of the digital nature of cyberspace and because of the universe of wanted and unwanted signals that propagate through cyberspace.

Spam is an unwanted cyber-signal.

Most anti-spam legal cases have so far viewed spam as a trespass to chattels—as so-called *virtual* trespass. The tort of trespass to chattels or personal property also dates back to medieval times in England. It is an intentional interference with one's personal property without consent or without some form of privilege. It usually requires some actual harm to one's personal property whereas trespass to land does not.

The original rationale for the virtual-trespass view was that spam interferes with servers and other hardware devices that store and process e-mail: "When an actor 'touches' another's web site or uses another's equipment to communicate in a way that interferes with, or affects, the owner's use, that actor has trespassed."[13] But few e-mail users can show any real monetary cost from deleting spam from their in-box. Most users pay no fee or they pay a flat monthly fee for Internet access. And most Internet service providers include spam filters as part of their service.

Private nuisance seems a more accurate description of spam— especially if we view spam as a form of cyber-noise.

Spam interferes with one's use and enjoyment of one's virtual space or address site in cyberspace. Again spam messages have grown exponentially with time. A steady stream of such unwanted e-mail noise signals should count as a substantial interference with one's virtual "land" or small chunk of cyberspace. And some legal scholars have argued for the land view when they accept that spam and similar unwanted Internet signals are trespass but argue that they are a form of virtual trespass to land rather than to chattels or personal property. We use Internet *addresses* to locate persons or firms in cyberspace much as we use physical-location addresses in society.[14]

So a steady spam bombardment can count as a private nuisance because it is a substantial and unwanted interference with one's virtual property—at least if a court "balances" the costs and benefits involved and finds that the interference is "unreasonable." The high transactions costs involved in locating and dealing with spam sources further favor nuisance over trespass per the above discussion of the Coase theorem.

At least one legal theorist has argued along the same lines that spam fits the elements of private nuisance better than it fits the elements of trespass to chattels: "A form of cyberspace nuisance claim, containing a healthy dose of real property doctrine, might better accommodate the peculiar calculus of benefits and harms in cyberspace."[15] Legal scholar Adam Mossoff has also challenged the virtual-trespass approach to spam but argued for a nuisance approach based more on interfering with chattels or personal property than on interfering with some sort of virtual land: "If injuries suffered to automobiles, lumber, crops, cattle, as well as general annoyance injuries caused by unsightly junk cars or excessive late-night telephone calls, are sufficient for a successful nuisance claim, then so are injuries suffered to computers and business operations by e-mail."[16]

Spam e-mails also look a lot like unwanted phone calls. Courts have had no trouble viewing such calls as a private nuisance as mentioned above. And Title 47 of the federal code outlaws sending unwanted advertisements by fax: "It shall be unlawful for any person within the United States . . . to use any telephone facsimile machine, computer, or other device to send an unsolicited advertisement to a telephone facsimile machine" (47 U.S.C. § 227(b)(1)(C)). Still plaintiffs may ultimately have to form a certified class in a class action to get a court to impose an injunction on spammers under either the legal theory of trespass or of nuisance.

Congress has made it somewhat easier for Internet service providers to sue spammers. Congress used its power to regulate interstate commerce to pass the CAN SPAM law in 2003. This law allows civil and criminal penalties for spam advertising that is fraudulent or deceptive. It requires that spammers give a "clear and conspicuous identification that [the] message is an advertisement or solicitation" or face fines or prison (§ 5 (a)(1) of 16 C.F.R. Part 316). California and other states have passed similar anti-spam laws but the federal law preempted them since in general federal law trumps state law. Microsoft and America Online and other Internet service providers used the new federal spam law in 2004 to bring common legal actions in federal courts against spammers. But anti-spam laws face the same kind of free-speech challenges that have defeated related measures to restrict unwanted phone calls from advertisers. And smart spammers will likely turn the slightest loopholes in such laws into digital floodgates for freshly crafted spam. They also tend to broadcast from foreign jurisdictions that help keep them beyond the reach of American or European law.

A good rule of thumb is that there will always be far more unwanted signals than wanted signals—especially in cyberspace. Treating spam as a cyber-noise nuisance is one way to fight back.

CHAPTER 3

Work-related hearing loss is one of the most common occupational diseases in the United States.

–Linda Rosenstock
former director: National Institute for Occupational
Safety and Health

Noises generated in industry, in the military services, and even in leisure activities may lead to permanent hearing impairment. Most industrial noises can be classified as either impact noises (drop hammers, punch presses, etc.) or as steady noise (diesel engines, lathes, etc.). The acoustic energy in noise is seldom distributed uniformly among the component frequencies. Thus piston engines and pit furnaces produce mainly low-frequency noise, whereas pneumatic hammers and high-speed saws produce predominately middle- and high-frequency noises.

–Linda Luxon
"The Clinical Diagnosis of Noise-induced Hearing Loss"
Biological Effects of Noise

As a musician, I learned early in my career the importance of healthy hearing, although many of my friends didn't realize the damage until it was too late.

> —Pat Benatar
>
> national spokesperson for Hearing Education
> and Awareness for Rockers (H.E.A.R.)

For soft but audible stimuli, an optimal amount of "prosthetic" noise significantly improves sensitivity to envelope modulation in cochlear implant listeners.

> —Monita Chatterjee and Mark E. Robert
> "Noise Enhances Modulation Sensitivity in Cochlear
> Implant Listeners: Stochastic Resonance in a
> Prosthetic Sensory System?"
> *Journal of the Association for Research in Otolaryngology,*
> vol. 2, no. 2, June 2001

Intermittent noise is one of the most frequent causes of sleep disorder. Intermittent noise can disturb sleep not only through subjective annoyance (causing awakening) but it can also lead to fundamental changes in the sleep structure without waking the individual. In this connection, fragmentation of the sleep structure, the reduction of rapid eye movement (REM) sleep, disturbance of mental and emotional processes and a longer period in shallow sleep are important. Intermittent noise, like flight noise, forces the sleeping person to repeatedly restart his sleep cycle, which requires high energy.

> —C. Mashke
> "Noise-induced Sleep Disturbance, Stress Reactions,
> and Health Effects"
> *Biological Effects of Noise*

Manatees may be least able to hear the propellers of boats that have slowed down in compliance with boat speed regulations intended to reduce collisions. Such noise often fails to rise above the noisy background in manatee habitats until the boat is literally on top of the manatee.

—Edmund R. Gerstein

"Manatees, Bioacoustics, and Boats"

American Scientist, vol. 90, April 2002

*We have found that urban great tits (*Parus major*) at noisy locations sing with a higher minimum frequency, thereby preventing their songs from being masked to some extent by the predominantly low-frequency noise. Anthropogenic noise could affect breeding opportunities and contribute to a decline in species density and diversity.*

—Hans Slabbekoorn and Margriet Peet

"Birds Sing at a Higher Pitch in Urban Noise"

Nature, vol. 424, 17 July 2003

How much noise does it take to damage the human ear?

Enough energy can damage or destroy any system. No car or human body can withstand the kinetic energy of a high-speed head-on collision. The sudden impulse of mechanical energy crumples metal and crushes bone and soft tissue. Heat energy greater than 6,000 degrees kelvin can melt a diamond. Great cosmological gobs of energy can in theory fold the space-time continuum and bring point *A* into direct contact with distant point *B*.

So it is no surprise that enough noise energy can damage something as delicate as the inner ear. A 1998 article in the *Journal of the American Medical Association* found that about 15% of American teenagers have permanent ear damage.[1] This damage presumably comes from listening to loud music—energetic signals to the teenagers but noise signals to many others. The type of music does not matter: Studies have shown that playing music in a symphony orchestra also causes hearing loss.[2] What matters is how much energy impinges on the inner ear and for how long.

3.1. NOISE-INDUCED HEARING LOSS IS A COMMON HEALTH HAZARD

Noise-induced hearing loss is so serious and widespread that it now has its own acronym in the medical and workplace literature: NIHL.

NIHL is a much bigger health problem than most people realize. The National Institute for Occupational Safety and Health reports that NIHL is one of the most common occupational illnesses and that work environments expose about 30 million Americans to NIHL noise hazards.[3] About 10 million of these persons have some permanent hearing loss from noise. A full 90% of miners can expect NIHL hearing loss at age 52 compared with only 9% of the general population. And the trend is toward even greater loudness: Reported hearing problems increased 26% during the period from 1971 to 1990 for those aged 18 to 44. Governments such as Sweden and many others regulate the noise levels at construction sites and other workplaces that tend to have high levels of NIHL. The U.S. Department of Labor sets occupational noise standards through the Occupational Safety and Health Administration (Section 1910.95 of its compliance code). These federal noise regulations apply only when workplace noise levels exceed some predefined thresholds.

We measure sound energy in decibels or the base-10 logarithm of a ratio of sound pressure.[4] The name "decibel" itself refers to one-tenth the unit of sound intensity named after telephone pioneer Alexander Graham Bell. The logarithmic definition of a decibel means that arithmetical changes produce geometric results: Adding 20 decibels multiplies the sound pressure by 10. So 60 decibels of normal human speech has a sound pressure a thousand times greater than the faint reference sound pressure that defines the scale at 0 decibels. Loud rock music at 120 decibels has a sound pressure a million times greater than the reference pressure. A deafening rocket engine at 180 decibels has a sound pressure a billion times greater.

The logarithmic structure of decibels reflects the so-called Weber-Fechner law of psychophysics that governs our hearing. This empirical "law" or trend states that a tenfold increase in the amplitude of a sound stimulus corresponds to what we perceive as the minimal incremental increase in loudness. Psychophysicists call one decibel the "just noticeable difference" in sound intensity.

Scientists model a pure tone or sound signal itself as an oscillating sine wave because sound gently compresses and expands air many times per second in a periodic fashion. Using sinusoids also makes the math easier. Science fiction films often implicitly get the sine-wave part right but the air part wrong when something blows up in space and there is any sound at all. This is but one way that sound energy differs from electromagnetic energy such as light. Sound cannot travel through the vacuum of space for the same reason that you cannot swim in an empty swimming pool.

Western music builds great sonic tapestries based on these sinusoids. The A note on a piano above middle C undulates 440 times per second. One second of orchestral film music can involve 50 different such notes and their characteristic overtones and timbres. The nineteenth-century physicist and physician Hermann Helmholtz propounded the influential thesis that music differs from noise precisely because Western music consists of superimposed sinusoids and hence periodic sound while much noise lacks periodicity: "The sensation of a musical tone is due to a rapid periodic motion of the sonorous body. The sensation of a noise is due to non-periodic motions."[5] We do tend to perceive some periodic sound as music. But we can at the same time hear someone else's sinusoid-based music as interfering noise. So Helmholtz's sinusoidal distinction between music and noise does not even hold for highly stylized Western music.

The human ear and auditory cortex can detect and process a wide range of sound intensities. We can hear sounds softer than a whisper at little more than 1 decibel. Normal conversation takes place at about 60 decibels. The threshold of pain starts for most people somewhere between 120 and 130 decibels. The table below lists some common sounds and their typical decibel levels.

DECIBELS	SOUND TYPE
0	Threshold of hearing
10	Human breathing
20	Rustling of leaves
30	Whisper
40	Residential area at night
50	Quiet restaurant
60	Normal talking
70	Busy traffic
80	Vacuum cleaner
90	Loud factory noise
100	Jackhammer
110	Motorcycle muffler
120	Threshold of pain
130	Rock concert

140	Stereocilia damage
150	Jet engine
160	12-gauge shotgun
170	Humpback whale song in water
180	Rocket engine

Table 3.1: Approximate decibel levels of common sounds.

Hearing loss can still occur at nonpainful decibel levels if the exposure lasts long enough. Two straight hours of 90-decibel noise can cause permanent hearing loss or NIHL. This accounts for NIHL in many factory workers. Parts of the manned Space Station emitted damaging noise of 70 decibels or more until astronauts installed sound-absorbing padding.[6] A more common risk of NIHL comes from spending just one minute at the 130 or so decibels near the stage at a rock concert. This explains why so many rockers are hard of hearing. Rock star Pat Benatar is the national spokesperson for H.E.A.R. or Hearing Education and Awareness for Rockers. She uses this forum to warn fellow musicians and baby boomers about the NIHL dangers of loud music.

There is even a market for intentional infliction of noise distress. Engineers at San Diego's American Technology Corporation have produced a "noise gun" that shoots focused "sonic bullets" made from the sound of a baby crying played backwards—but at 140 decibels. The acoustical weapon uses a flat panel of sound amplifiers and can stop a charging assailant or force a bank robber to retreat.[7]

The European Union has led the world in decibel-based noise consciousness by drawing the first noise maps for cities. These maps use colors to show the noise intensity in decibels. They also are three-dimensional. The maps can aid builders and politicians and can presumably affect real estate prices. The editors of *Nature* rightly lauded the new noise maps and observed that noise harm goes far beyond hearing loss: "Some 80 million Europeans already suffer from noise above 65 decibels, enough to cause lost sleep, stress, high blood pressure and even heart attacks, and a further 170 million endure levels that are simply annoying."[8]

Still hearing loss remains by far the most common form of noise harm.

3.2. NOISE CAN DAMAGE THE INNER EAR'S FREQUENCY DETECTORS

A single energetic noise pulse can cause deafness. It takes only one handgun or shotgun blast near an ear to cause permanent hearing loss. The cannons of the Civil War and First World War must have produced a great deal of deafness in those who fired them without adequate ear protection. Modern firecrackers can produce the same effect if they explode close enough to the ear. Even the whining roar of a motorcycle can cause deafness if the ear is too close to the muffler. A car's air bag can cause NIHL when it bursts open. Again less energetic sound can cause comparable damage if the sound lasts long enough. That is a good reason to turn down the volume when wearing a headset or not to wear a headset at all.

The energy from a loud noise impulse can damage the fine hairs in the inner ear. Each inner ear has about 16,000 or so of these hair cells. They do not undergo cell division or otherwise grow back if damaged—though someday stem cells or other biotechnology may induce such growth. Loss of hair cells is the most common cause of deafness.

The hair cells themselves act as frequency detectors. Different hairs respond to different frequencies of sound or noise as the signal energy works its way as a mechanical sound wave from the vibrating eardrum through fluid canals back up to the curled-up cochlea in the inner ear.

The length of a hair cell tells what frequency it codes for in an inverse relationship between length and frequency. The shortest hairs detect the highest-frequency signals (about 20,000 hertz or cycles per second) while the longest hairs detect the lowest-frequency signals (about 20 hertz). The hair cells lie on the part of the cochlea called the organ of Corti. Scientists group the hairs as so-called inner hairs and outer hairs. The inner hairs respond to given frequencies while the outer hairs help sharpen the frequency response of the inner hairs. Loud noise can bend or break the hairs. This can result in deafness for the specific frequency that such hair detects. It can also produce false electrical signals that result in the persistent ringing of tinnitus.

We can take the analysis of noise damage to a level deeper. Loud noise can also damage the hair cell's mechanism that produces the electrical signals that travel along the auditory nerve from the inner ear to the brain. The inner-ear hairs convert sound energy into electrical signals by using the energy to open and close ion channels. Positively charged potassium ions (K^+) flow through the protein-based channels and create the electrical impulses. A single hair cell may contain 100 ion channels. The hair cells themselves often exist in clumps of 100 or so.[9]

Loud noise can damage the gates that open and close the ion channels and so can reduce or eliminate the resulting electrical impulses. Loud noise that bends a hair may result in closing or blocking some of the ion channels. This resembles the chemical damage that antibiotics such as streptomycin can produce by blocking the ion channels. Other bending may keep the channels permanently open and so produce the false electrical signals of tinnitus.

Yet even here noise can help.

Small amounts of noise can improve hearing by helping the inner ear's hairs detect frequency signals. This is the so-called stochastic resonance effect that we saw briefly in chapter 1 and will see again in detail in chapter 6. Direct experimental evidence of this positive noise effect has come from Monita Chatterjee and her colleagues in their work with cochlear implants at the House Ear Institute in Los Angeles. Other researchers have found mathematical and simulation evidence that cochlear hairs should benefit from small amounts of noise.[10]

Cochlear implants try to detect the same frequency signals that healthy inner-ear hairs detect. The user may wear a pocket-sized signal processing unit that transmits detected signals to small antennas on eyeglass frames or other structures. Then the device sends electrical signals through electrodes to stimulate auditory-nerve neurons and thereby approximate the input-output effects of healthy hair cells. Chatterjee and her colleagues have shown that an optimal amount of noise boosts the hearing of implant wearers when they listen to pure tones such as one might hear from a vibrating tuning fork. It is an open question whether there is an optimal amount of noise that helps implant wearers when they listen to human speech.

Cell phones offer a related opportunity for a noise boost but one that could harm hearing. A Swedish study found that cell phone users had an increased risk of growing acoustical neuromas or benign tumors on their acoustic nerve (the eighth cranial nerve) if they used an analog cell phone for at least 10 years. (Lesser use produced no detectable effect.) The tumor risk was four times normal on the side where the patient held the cell phone to his ear and normal on the other side.[11] The 2004 study looked at only analog cell phones because digital cell phones had not been in sufficient use for ten years. The study did not determine whether the tumor growth came from the signal structure (or carrier frequency) of the transmitted sounds or from the cell phone's power source. Either source of faint energy could count as a noise source that could in principle help trigger an on-off thresholdlike mutation at the site of the gene on chromosome 22 that apparently controls neurofibromatosis and the growth of the Schwann cells that surround nerves. The explosive growth in cell phone use should in time provide ample data to test such noise hypotheses if cell phones in fact increase tumor risk.

3.3. NOISE INCREASES STRESS

Noise can cause stress. Ringing phones and shouting children may merely annoy a person who strolls through a grocery store or a public park. The same noise can ruin the experience of watching a film in a theater or in a living room. Such noise can also disrupt or even prevent sleep—and sleep loss promotes stress and a variety of health problems from increased blood pressure to decreased immune response. A study of German children who live near Munich airports found that airport noise impaired the children's long-term memory and their ability to read.[12]

The steady rise in noise pollution correlates to some extent with the rise in sleep deprivation. *Newsweek* reported a startling statistic based on the European Union's 3-D noise maps: "A single noisy motor scooter driving through Paris in the middle of the night can wake up as many as 200,000 people."[13] Noise-induced sleep loss only compounds the sleep deprivation of so many busy people in the modern workplace—people who get through their overscheduled days or nights on too little sleep and too much caffeine and related stimulants.

Our noise sensitivity is likely hardwired because we did not evolve to sleep or wake in high-decibel noise environments. Our mammalian brains and endocrine systems evolved in low-decibel environments over millions of years. High-decibel sounds were stressors such as screams or animal roars or even thunderclaps. These were exceptional signals in exceptional circumstances that could easily affect survival because the associated actions could damage tissue. So selection pressure favored those hairy and hairless hominids who best correlated loud noises with stress hormones such as cortisol and adrenaline and with fight-or-flight behavioral responses. That genetically encoded noise sensitivity helped keep us from going extinct. The modern cost is that more and more people live a life full of noise-induced stress—even before the invention of the iPod and ever more powerful car stereo speakers. A hunter-gatherer's sensitivity to high-decibel noise does not promote Zen calm or good digestion on a Monday morning while walking against the sidewalk crowd in New York City.

3.4. NOISE CAN HARM SIMPLER ANIMALS

The evolutionary argument above suggests that unnatural noise should produce similar stress in lesser animals. Growing evidence supports this conjecture.

Low-frequency sonar and other man-made signals appear to harm whales and other marine mammals that rely on their hearing to detect prey sound signals and to communicate with one another and to avoid predators. Sound travels up to five times faster in water than in air. Sonar signals from military or commercial vessels can propagate great distances and can produce energetic and unnatural noise interference. Even slow-going recreational boats can produce low-frequency noise that confuses endangered Florida manatees and so increases the chance that one of these large sea creatures will run into a boat propeller. This has led to increasing calls for boats to install signaling devices on their bows to help these would-be mermaids avoid maiming or fatal collisions.[14]

The U.S. Navy's new low-frequency-active sonar also appears to harm whales. The Navy uses this powerful low-frequency sonar to detect new ultra-quiet submarines at ever longer ranges. Geophysicists also want to use this sonar to help study the geological structure and history of the ocean floor. Scientists already bounce low-frequency signals between California and Hawaii to measure large-scale changes in North Pacific water temperature as part of the Acoustic Thermometry of Ocean Climate project. The 1972 Marine Mammal Protection Act permits these and other noise experiments in the oceans and requires noise-impact studies.

Navy experiments with low-frequency sonar and mid-frequency sonar near the Bahamas have coincided with an increased incidence of beaked whales getting stranded on nearby beaches. Some of the beached whales bled from the ears and showed CAT-scan signs of trauma in ear and brain tissue.[15]

The Navy has disputed environmentalists' claims that the new low-frequency sonar harms whales or other marine life because it uses lower-frequency signals than the signals apparently involved in the sea-mammal mishaps. It is far from certain that low-frequency signals cause direct harm to marine mammals. And statistics always come with the caveat that correlation suggests causation but correlation does not itself prove causation. Other variables could also have intervened. Lightning and landslides and other natural events can also produce powerful low-frequency signals in oceans.

Still there is some evidence of a causal relationship. One research team performed autopsies on 14 beached beaked whales near the Canary Islands where the Spanish navy had just conducted exercises with sonar. The team found circumstantial tissue evidence that the whales had died from decompression damage to their livers and kidneys. The whales beached themselves just four hours after the naval exercises began. The sonar noise may spook underwater whales and make them head too quickly to the surface. The researchers speculated that such rapid ascent may produce humanlike decompression sickness or "the bends" in the whales.[16]

Sonar also increases the length of whale songs. Biologists cooperated with the U.S. Navy in a unique controlled experiment at sea. The scientists wanted to see if the Navy's new low-frequency sonar affected the singing of male humpback whales. It did: The sonar increased the average length of whale songs by 29%. The song lengths returned to normal in the absence of the sonar.[17] The whales sing their eerie and far-reaching songs to help navigate the world's oceans and to find and court potential mates. They presumably sang longer in the presence of the sonar noise to maintain a comparable signal-to-noise ratio amid the man-made acoustical interference. It is possible that small amounts of sonar noise could shorten whale songs or even help beaked whales better detect sharks and prey sources. So far there have been no such oceangoing experiments to detect such stochastic resonance effects.

More common noise can disturb the environment in subtle ways. Biologists have found evidence that urban noise affects the communication signals of wild songbirds. They studied the frequency structure of the songs from 32 male urban great tits (*Parus major*) in the Dutch city of Leiden. The yellow-bellied birds sang at a higher frequency to overcome the low-frequency background noise of the city.[18] Not all creatures may be as plastic in adapting to new high-decibel noise environments. This may be one more factor working against the survival and diversity of species.

Modern man creates the noise pollution that makes the love-seeking whales and songbirds sing longer and louder just to hear one another. Yet no one may suffer more from this potentially deafening nuisance than the hominid who creates it. We sleep less and release more stress hormones as we respond to a noisy world of ever higher decibels and ever richer signal complexity.

Noise is a by-product of modern life. It promises only to get worse as cities expand and as more people use more gadgets. Often such noise is a mere petty nuisance that we can avoid by closing a door or shutting a window. Sometimes it affects the public health as with that lone and loud motorcycle waking up thousands of Parisians.

Either way the first step is accurate measurement: Every city needs to publish its own noise map.[19]

Each vibrating electron constitutes a tiny current. *The net effect of billions of random electron currents in a resistive material produces the phenomenon of* thermal noise. *Since we cannot build an electrical system without electrons and resistance, thermal noise is inevitable—like death and taxes.*

  –Bruce Carlson and David Gisser

    Electrical Engineering

In this paper it will be shown that according to the molecular-kinetic theory of heat, bodies of microscopically-visible size suspended in a liquid will perform movements of such magnitude that they can be easily observed in a microscope on account of the molecular motions of heat. It is possible that the movements to be discussed here are identical with the so-called "Brownian molecular motion."

  –Albert Einstein

    "Investigations on the Theory of the Brownian Movement"

    Annals of Physics, 1905

Most engineering systems in communication, control, and signal processing are developed under the often erroneous assumption that the interfering noise is Gaussian. Many physical environments are more accurately modeled as impulsive, characterized by heavy-tailed non-Gaussian distributions. The performances of systems developed under the assumption of Gaussian noise can be severely degraded by the non-Gaussian noise due to potent deviation from normality in the tails.

–Seong Rag Kim and Adam Efron

"Adaptive Robust Impulse Noise Filtering"

IEEE Transactions on Signal Processing, vol. 43, no. 8, August 1995

The elementary atomic process underlying the flicker effect is the appearance of an individual foreign atom or molecule in the surface of the cathode, changing the ability of the surface to emit electrons so long as the foreign atom remains.

–Walter Schottky

"Small-Shot Effect and Flicker Effect"

Physical Review, 1926

A single lightning stroke radiates considerable radio frequency noise power. At any one moment there are an average of 1,800 thunderstorms in progress in different parts of the world. From all these storms about 100 lightning flashes take place every second. The combined effect of all lightning strokes gives rise to a noise spectrum which is especially large at broadcast and shortwave radio frequencies.

–Merrill L. Skolnik

Introduction to Radar Systems

Early SETI projects concentrated on listening for electromagnetic signals in the centimeter waveband between 3 and 60 centimeters. The reasons for this choice are purely practical: It is in this region of the radio spectrum that background noise from the Galaxy, the Earth's atmosphere, and the receiving equipment is lowest.

–T. L. Wilson
"The Search for Extraterrestrial Intelligence"
Nature, vol. 409, 22 February 2001

It should always be remembered that what is real *is some inexplicable wiggly line in a spectrum—and what is a figment of our imagination is a* $^2T_2 \rightarrow {}^2A_2$ *transition.*

–Daniel C. Harris and Michael D. Bertolucci
Symmetry and Spectroscopy

Black-noise phenomena govern natural and unnatural catastrophes like floods, droughts, bear markets, and various outrageous outages, such as those of electrical power. Because of their black spectra, such disasters often come in clusters.

–Manfred Schroeder
Fractals, Chaos, Power Laws

Popcorn noise, also called burst noise, was first discovered in semiconductor diodes and has recently reappeared in integrated circuits. If burst noise is amplified and fed into a loudspeaker, it sounds like corn popping, with thermal noise providing a background frying sound—thus the name popcorn noise.

–Henry Ott
Noise Reduction Techniques in Electronic Systems

Not all systems crackle. Some respond to external forces with many similar-sized small events (popcorn popping as it is heated). Others give way in one single event (chalk snapping as it is stressed). Crackling noise is between these two limits.

–James P. Sethna, Karin A. Dahmen,
and Christopher R. Myers
"Crackling Noise"
Nature, vol. 410, 8 March 2001

Apart from intrinsic noise sources at the level of an individual neuron there are also sources of noise that are due to signal transmission and network effects. Synaptic transmission failures, for instance, seem to impose a substantial limitation within a neuronal network.

–Wulfram Gerstner and Werner Kistler
Spiking Neuron Models: Single Neurons, Population, Plasticity

As the universe expanded from its hot, dense origins, all the structures seen in the universe today formed under the action of gravity on initial nearly scale-invariant adiabatic Gaussian density fluctuations.

–A. C. S. Readhead et al.
"Polarization Observations with the Cosmic
Background Imager"
Science, vol. 306, 29 October 2004

If one makes the hypothesis that the maximum amount of information which can be lost down a hole of a given mass, angular momentum, and charge is finite, it follows that one can associate an entropy with the hole and deduce that it must emit thermal radiation at some finite nonzero temperature.

–Stephen W. Hawking
"Black Holes and Thermodynamics"
Physical Review D, January 1976

For the theory that fits our data, the Universe will expand forever.
–Wilkinson Microwave Anisotropy Probe (WMAP) Web site

Is the universe noise?

That question is not as strange as it sounds. Noise is an unwanted signal. A signal is anything that conveys information or ultimately anything that has energy. The universe consists of a great deal of energy. Indeed a working definition of the universe is all energy anywhere ever. So the answer turns on how one defines what it means to be wanted and by whom.

Suppose you listen to music or a talk-show host's voice on the radio. Then all other signals and forms of energy amount to distractions or interference to varying degree—even if they are faint cosmic rays from distant galaxies. But the signal-noise duality lets a sincere pantheist counter that he loves or wants God and that God just is the entire universe but spelled with fewer letters. So to him the universe is not noise but one big wanted signal. There is no definitive answer here because of the subjective nature of wanted versus unwanted signals in the definition of noise.

A different answer emerges from the extensive science and engineering of noise: The universe emits noise.

The universe is noisy on all scales even if the universe itself is not noise. The fading noise of the ancient big bang explosion fills the cosmos. It still gently hisses and crackles all around us in the form of junk microwave radiation. Measurement and device noise attaches to all our measurements and devices. Our warm brains give off thermal noise while such thermal and other noise types infest the fine electrical circuitry of the neural networks that make up our brains. And peering down into the quantum world reveals noise fluctuations in the ultimate substrates of matter. Even black holes are noisy.

The universe is a noisy place and its noise will only grow as it expands forever.

Noise is destiny because the universe will end in noise.

4.1. WHITE NOISE IS INDEPENDENT IN TIME AND HAS A FLAT SPECTRUM—AND SO IS PHYSICALLY IMPOSSIBLE

Any technical discussion of noise begins with white noise because white noise is pure or ideal noise. White noise serves as the gold standard of noise. Scientists and engineers have explored hundreds of other noise types but most of these deviate from white noise in some specific way.

White noise is noisy because it has a wide and flat band of frequencies if one looks at its spectrum. This reflects the common working definition of noise as a so-called wideband signal. Good signals or wanted signals concentrate their energy on a comparatively narrow band of the frequency spectrum. Hence good signals tend to be so-called narrowband signals at least relative to the wide band of white noise. White noise is so noisy because its spectrum is as wide as possible—it runs the whole infinite length of the frequency spectrum itself. So pure or ideal white noise exists only as a mathematical abstraction. It cannot exist physically because it would require infinite energy.

The same fate holds for many mathematical abstractions. A common and related example is a sine wave that undulates in time and that we approximate with the ringing of a tuning fork. A true sine wave technically wiggles all the way out to positive infinity and all the eternal way back in time to negative infinity. Such infinite repeating waves would also have infinite energy and so do not exist. Real wiggling waves are actually *wavelets* that wiggle only over a finite stretch of time and so have finite energy. White noise is so idealized that most forms of it do not even technically exist in the artificial world of mathematics (because such white noise is the pseudovelocity or instantaneous rate of change of a wildly jiggling Brownian motion).[1]

White noise sounds like the hiss and pop of radio static or the clatter of raindrops hitting a roof or sidewalk. Energetic versions can sound like a waterfall or the crash of white-water rapids. Many of us sleep at night with a fan blowing at us to approximate the sound of a gentle white noise and to mask background noise inside and outside the room. Yet too much white noise may harm the auditory portions of the brain—at least if you are a baby rat.[2]

But just what is white about white noise? How can sound have color?

Sound itself has no color. Sound conveys the energy of air pressure over time. So it is a time signal or a signal that varies in time. We saw in the last chapter that the decibel measures the intensity or energy level of a sound signal and in general of a time signal. The color comes from an analogy with the frequency representation of the time signal. The spectrum of pure white noise looks like the flat spectrum of pure white light.

Pure white light consists of all colors in the visible spectrum equally present in a light signal. So white light is a form of broadband noise—and so *daylight is noise.*

Each color has its own narrow band of frequencies and its own wavelength from wave crest to wave crest. The speed of a wave equals its frequency times its wavelength. This gives a special relationship for light because the speed of light is constant in a vacuum. Light in empty space travels at about 300,000 kilometers per second or about 186,000 miles per second. The speed of light is also constant but slower in other transmission media such as air or water or honey.

A glass prism splits white light into light of different colors and thus splits it into different frequencies and wavelengths. Red light has a longer wavelength than blue light. So red light has a proportionally lower frequency than blue light because red light and blue light both travel at the same speed. Red light has a wavelength of about 700 nanometers and thus has a staggering frequency of oscillation of almost 429 terahertz—about 429 *trillion* oscillations or cycles per second. Blue light has a shorter wavelength of about 450 nanometers with a corresponding increase in frequency.

These tiny distinctions in wavelength have profound effects in nature. Consider a green grape leaf or the red color of our blood. The grape leaf and the blood consist of atoms that formed billions of years ago when nearby supermassive stars exploded and died. The grape leaf appears green because of its chlorophyll molecules. It will later appear yellow or brown as it loses these chlorophyll molecules in the fall.[3] Four nitrogen atoms surround a lone magnesium atom inside a chlorophyll molecule—just as they surround a lone iron atom inside a hemoglobin molecule in blood. This ring structure allows chlorophyll to act as an antenna for red and blue light in the white light that the sun freely disperses in all directions from its surface. The chlorophyll molecule absorbs the red and blue light in the white light. Then the grape plant uses this absorbed light to energize the production of sugar or glucose in photosynthesis—sugar that we may later taste in the squish of a sweet table grape or in the sip of a bittersweet wine. But the chlorophyll molecule reflects the intermediate green light from the white light it receives. The reflected green light is a type of optical waste "by-catch" of the leaf's solar detectors. Some of the reflected green light waves strike our retina and stimulate our brain. This triggers neural action in a cascade of neural networks that in turn produce the pattern or illusion of green somewhere deep in our visual cortex toward the back of the skull. That is why we see grass as green and not as red or blue.

Again pure white noise has constant energy across an infinitely long swath of frequencies. So white noise has infinite energy if there is any height at all to the flat frequency spectrum because if we multiply infinity by any positive number then we still get infinity. We will see below (and in figure 4.6) that this need not be true for other types of "colored" noise that have a nonflat frequency spectrum that still runs over an infinite range of frequencies.

Visible white light has a reasonably flat spectrum across its small band of frequencies. Visible light itself takes up a minuscule mid-level portion of the entire electromagnetic spectrum. The full spectrum extends all the way down to the ultra-low-frequency signals that submarines use for communication and that they encounter as noise from surface ships. It extends all the way up to the superhigh frequency of energetic gamma rays that arise from atomic decay and that can kill cancer cells or cause radiation poisoning. The frequency oscillation of a gamma ray is on the order of a million trillion cycles per second. Gamma rays have so much energy because a core fact about quantum mechanics is that radiant energy is directly proportional to frequency (via Planck's constant). Even an approximately white electromagnetic signal would require tremendous energy because it would equally invoke all such wavelengths from the miles-long macrowaves of submarines to the smallest gamma waves on the order of a trillionth of an inch. Some solar flares do produce similar all-spectrum bursts of energy.

The flat spectrum of white noise conveys a remarkable and ideal property in the time domain: White noise is *independent in time*. Pure independence in time implies a pure flat frequency spectrum and vice versa. Real signals have nonflat spectra and thus the time signals correlate with one another to at least some degree.

Time independence explains the peculiar sound of white noise. Each hiss and pop in white noise is technically independent of the hiss and pop that preceded it in time and that follows it in time. This time independence is the *whiteness* property of white noise. Again it comes at the high price of physical impossibility because it would require a system with infinite energy and all known energy sources are finite. White noise in practice is often only a crude approximation to this ideal.

Here is the strange part: The time independence of white noise holds no matter how infinitesimally close a hiss is in time to the next hiss or pop.

That goes against much of daily experience. What we do or say or think in the next instant tends to look a lot like what we do or say or think in this fleeting instant. Human speech is so redundant that there is little change in a speech signal from instant to instant if you zoom in close enough. So you can often accurately predict the speech signal at the next instant in time if you know the signal at the current instant. That signal dependence lets engineers compress speech signals by throwing out much of the signal and just transmitting the parts that change. Many image compression schemes use the same idea to compress still pictures or moving images: Keep the independent or uncorrelated portions and drop much of the correlated portions. So it is impossible to accurately compress a pure white-noise speech signal or a fluttering white-noise TV image. The time independence of white noise makes each hiss and pop novel. Control engineers sometimes refer to white noise as an *innovations* process because each instant of it is unpredictable and hence fresh or new.

Time independence can produce a sound all its own. We hear the white noise as static if the hisses and pops occur several times per second as in radio static. The white-noise static becomes more of a sustained hissing sound as the frequency of the hisses and pops increases. But a white noise still results if the hiss-pop frequency slows to just one hiss or pop per second or even one per minute or day or year even though we no longer hear a hissing or static sound. The only requirement is that the hisses and pops be statistically independent of one another.

A discrete and slow version of time independence occurs if you go for a random walk while flipping a fair coin. Stand along a painted line or along a row of tiles on a tiled floor. Pull out a coin and flip it. Take one step to the left if the coin comes up tails. Take one step to the right if it comes up heads. Then keep flipping and stepping and you are going for a random walk in one dimension.

This random walk approximates the time independence of white noise because the outcome of one randomly flipped coin has nothing to do with the outcome of the next coin flip or with any other coin flip. Many a casino gambler denies this fact when he sees someone flip three coins in a row and get heads on all three flips and then wrongly concludes that the next flip is more likely to come up tails. The probability math says that there will be such "runs" of all tails or all heads and of all lengths. The catch is that the probability of a run tends to fall off exponentially with the length of the run. But this is only a tendency. The next coin-flip outcome always owes its heads or tails status to the "luck of the draw."

Now imagine shrinking the random walk. Keep flipping the coin and taking steps to the left or the right along a straight line. But imagine your step size getting smaller and smaller. Shrink the step size all the way down to the mathematical limit of an infinitesimal step size. Then your random walk along a straight line has just become a Brownian motion. Any two points are independent no matter how infinitesimally close they are. So knowing all the steps you took in your random past does not help at all in predicting what steps you will take in the future or even in the next instant.

Signal processors often soften the independence condition somewhat by saying that any two events are uncorrelated. This is a technicality because independent events are always uncorrelated but the converse is not true in general (though it is for the normal bell curve as in the first probability curve in figure 4.1). Still this reflects our statistical common sense as when we say that price inflation correlates with the money supply or that fatal car accidents correlate with high blood alcohol concentrations. There are statistical tests for independence called chi-square tests and sometimes these tests do show that two variables have little or no dependence or correlation. But dependence is the norm in a universe in which everything connects to everything else if only faintly and if only through gravitational attraction. The total time independence of white noise severs all such bonds.

4.2. THERE ARE INFINITELY MANY TYPES OF WHITE NOISE

White noise requires only that noise spikes be independent of one another in time or that they be statistically uncorrelated. Time independence places no limit on the probability structure of any given noise spike. A given noise spike can be small or large or be positive or negative. This lets us define an infinite variety of white noises based on the probability distribution that decides which step or sample to take next and with what probability. Figure 4.1 displays three such types of white noise and the corresponding probability curve or density function that governs the random but uncorrelated steps or spikes.

Figure 4.1 shows that there are at least three different types of white noise. How many types of white noise are there? There are infinitely many types and each has a flat frequency spectrum because all the noise spikes are independent of one another. Indeed there are as many types of white noise as there are real numbers on the real line because there are as many continuous probability curves or so-called probability density functions as there are real numbers. So it is not technically accurate to speak of white noise without stating the probability type. Still this did not keep author Don DeLillo from winning the National Book Award with his 1985 novel *White Noise*. The entertaining novel about coping with the fear of death helped give the

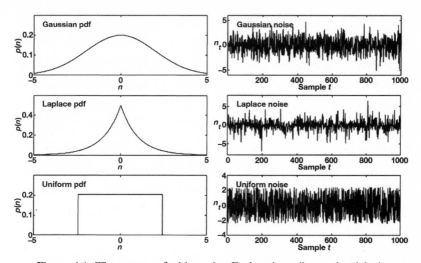

Figure 4.1: Three types of white noise. Each noise spike on the right is uncorrelated with every other noise spike in the time-series plot and so each noise type has in theory a flat frequency spectrum (not shown). The probability density function (pdf) on the left describes the probability distribution of the noise random variable n and so describes the probability that any particular noise spike will occur. The symmetry about zero of each probability curve implies that the noise random variables have zero mean. The Gaussian (normal) and Laplacian probability curves take their names from great European mathematicians and scientists of the early nineteenth century. The curves describe noise that is more likely to occur near zero. The uniform probability density function restricts the noise values to the interval $(-2, 2)$. But the noise values in that interval are equally likely to occur. The Gaussian or normal bell curve in the first figure is an example of what many people mean by *the* bell curve of probability as in the distribution of IQs or other standardized test scores.

term "white noise" some pop cachet among the literati. The 2005 horror film *White Noise* introduced the term to teens. The film suggested that ghosts hide signals in radio static.

Each probability curve describes the chance of getting a noise spike of a given magnitude. Each noise spike is a hiss or a pop. Note that the probability curves in figure 4.1 are symmetric about zero. So the left and right sides are mirror images of each other. This implies that the probability of getting a noise spike with a value near 1.5 is the same as getting one with a value near −1.5. The symmetry about zero means that on average the positive and negative noise spikes wash out to zero. The width of the curves describes the spread or dispersion about this average zero value. Fatter or wider curves describe experiments with less certainty because a wider range of noise spikes is more likely to occur. Wider bell curves also describe more energetic noise. Narrower probability curves concentrate the spike probability on a smaller range of values and correspond to less energetic noise.

Each probability curve has the key property that the total area beneath the curve equals exactly one. This means that it is certain that a noise spike will occur somewhere beneath the curve at any given moment. The different heights of the first two curves show the relative probabilities of the noise spikes and thus show that the most probable or most frequent spikes will tend to be near zero.

The second curve uses the Laplacian probability curve (named after the great French mathematician and scientist Pierre Laplace [1749–1827]) that engineers often apply to models of speech or image noise. The Laplacian probability curve concentrates its mass near zero. So most of its noise spikes are near zero in the simulation plot.

The flat probability curve in the third figure says that all the noise spikes are equally likely. This uniform distribution captures what most laymen mean by "randomness"–a random experiment where all outcomes are equally likely to occur as in drawing names from a hat or drawing ping-pong balls from a bingo tumbler. But randomness deals with the vastly more general case where a probability curve describes the unequal likelihood of occurrence of two or more outcomes.

The first probability curve in figure 4.1 is an example of the most famous probability curve in all of science and mathematics. It is a so-called normal or Gaussian bell curve named after the great German mathematician and scientist Johann Carl Friedrich Gauss (1777–1855).

Gauss either founded or profoundly contributed to every major branch of mathematics as well as physics and other fields. That helps explain why so many of us consider Gauss the greatest mathematician of all time and indeed one of the greatest intellects of all time.

Gauss's bell curve describes many physical and social processes as well as industrial processes. It describes the distribution of heights of a country's boys or men of a given age as well as the narrow spread about the mean in the weight of a box of cornflakes grabbed at random from a store shelf. The mean describes the location of the center of the bell curve along the horizontal axis. The mean is zero in figure 4.1. The variance measures the bell's width or the dispersion of random samples about the mean. Fatter curves have a larger variance and imply that there is less certainty about the mean than with a thinner curve. The square root of the variance is the standard deviation and we measure it too along the horizontal axis. This is a technical but important detail—we measure means and standard deviations on the same scale.

Standard deviations measure the statistical dispersion distance from the mean of the bell curve. About 68.2% of the area of a normal (Gaussian) bell curve falls within one standard deviation of either side of the mean. So there is a 68% chance that any given noise value will lie one standard deviation to the left or to the right of zero in the first curve in figure 4.1. About 95% of the area falls within two standard deviations of the mean. About 99.7% of the area falls within three standard deviations of the mean. And about 99.9999998% of the area falls within six standard deviations from the mean. Statisticians use the lowercase Greek symbol σ (sigma) to denote one standard deviation. This explains why some production managers and quality-control engineers talk about achieving *Six Sigma* product quality. This means that super-high-quality product lines should have (after some uncertainty rounding) only about 3.4 defective products out of every one million products that roll off the

assembly line. Six Sigma quality control may be more hype than fact
in many cases because most real-world bell curves have fatter tails
than Gauss's bell curve as we discuss below.

Gauss's bell curve is so common in science that when scientists
or engineers refer to white noise they almost always mean by default
Gaussian white noise. Communication engineers call this "wagon"
noise or AWGN–additive white Gaussian noise. This default model
assumes that that normal white noise corrupts a signal by adding to
it. Conventional radio and television broadcasting work with such
AWGN models of noise corruption.

Important exceptions include image models that describe the
blurring of an image or photo by assuming that image noise corrupts
the image by multiplying the image signal rather than by adding to
it. This leads to a whole family of deblurring algorithms that try to
reverse the process. *Speckle noise* is another example of multiplicative
noise. Lasers and ultrasonic images and some types of radar can
produce unwanted spots or speckles because of echo effects.

Still additive white Gaussian noise remains the overwhelming favorite noise model in science and engineering. It has a relatively simple
mathematical description and it often gives a reasonably good approximation of many types of real noise. Both the simplicity and the model
accuracy stem from the nature of the ubiquitous Gaussian bell curve.

The Gaussian bell curve lies at the heart of the most famous result in probability and statistics–the central limit theorem. This result is what people likely refer to when they talk about the fictitious
"law of averages." There is no law of averages. Instead there are laws
of large numbers that tell us that if we take enough samples from a
population then the sample averages come arbitrarily close to the
unknown population average. These laws underlie reported economic statistics such as unemployment rates and income levels for
the entire population because statisticians may have sampled only a
few thousand workers or wage earners out of a total population of

many millions. Pollsters and scientists use such results every day even though such averaging can fail catastrophically in some cases.[4] The endnotes discuss this notoriously tricky subject in detail.

The central limit theorem differs from laws of large numbers because random variables vary and so they differ from constants such as population means. The central limit theorem says that certain independent random effects converge not to a constant population value such as the mean rate of unemployment but rather they converge to a random variable that has its own Gaussian bell-curve description.[5] IQs obey a Gaussian curve with a center or mean of about 100. Students who get over 1520 on their combined SAT scores can join the Triple Nine Society because they allegedly have an IQ in the far-right 99.9% tail of the Gaussian IQ bell curve.

The central limit theorem lets statisticians form confidence intervals about such averages. That lets them test formal statistical hypotheses such as whether a cluster of student test scores arose from chance or from cheating. The central limit theorem similarly accounts for both the many political polls we see during election season and especially for the obligatory footnote that reminds us that the poll has a margin of error of plus or minus 3%.[6] This usually means that the pollster got responses from at least 1,068 subjects.

The central limit theorem also explains how a random walk converges to a Brownian motion as the step size of the random walk gets smaller and smaller. Such Brownian jiggle of colliding and vibrating atomic particles produces the thermal noise that we discuss below.

Still this is just the tip of the noise iceberg. We next look at two broad categories of noise that further generalize ideal white noise—though in these and other cases we only hint at the vast literature on these subjects. The first category consists of impulsive noise that can still be white but where the white noise is even more random or erratic than the relatively well-behaved types of white noise in figure 4.1. The second category generalizes in a different direction. It drops the assumption of a flat frequency spectrum and thus allows the noise to have statistical color.

4.3. MOST NOISE IS IMPULSIVE

Many scientists who work not just with noise but with probability make a common mistake: They assume that a bell curve is automatically Gauss's bell curve. Empirical tests with real data can often show that such an assumption is false. The result can be a noise model that grossly misrepresents the real noise pattern. It also favors a limited view of what counts as normal versus non-normal or abnormal behavior. This assumption is especially troubling when applied to human behavior. It can also lead one to dismiss extreme data as error when in fact the data is part of a pattern.

The reason for the mistake is that all too many scientists simply do not know that there are infinitely many different *types* of bell curves. So they do not look for these bell curves and thus they do not statistically test for them. The deeper problem stems from the pedagogical fact that thick-tailed bell curves get little or no mention in the basic probability texts that we still tend to use to train scientists and engineers. Statistics books for medicine and the social sciences tend to be even worse in this regard.

Another problem is that the competing bell curves look pretty much the same to the untrained eye. They all have a rough bell shape. And they all involve a generalized form of the central limit that arises when several independent effects combine. But the bell curves differ categorically in the probability world they describe.

A stark example of the assumption that all bell curves are Gauss's bell curve is the title of the controversial 1994 book about IQ and race called *The Bell Curve*. The definite article "the" in the title asserts a unique status about Gauss's bell curve that does not hold in the real world or in the ideal world of mathematics. It is even worse for the common management hype that surrounds "Six Sigma" concepts because most bell curves have such thick tails that they don't have a "sigma" or standard deviation at all.

Bell curves don't differ that much in their bells. They differ in their *tails*. The tails describe how frequently rare events occur. They describe whether rare events really are so rare. This leads to the saying that the devil is in the tails.

Consider first the case of the Cauchy bell curve and the Cauchy white noise that it spawns. Augustin Cauchy (1789–1857) was a titanic French scientist and mathematician. He was Gauss's contemporary and equal in many ways in terms of the breadth and depth of his technical achievements. Much of what mathematicians call "analysis" depends on concepts and theorems that still bear his name. Cauchy's bell curve has thicker tails than Gauss's curve does. This leads to white noise that can be wildly *impulsive* as figure 4.2 shows:

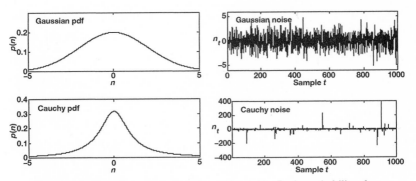

Figure 4.2: Gaussian versus Cauchy white noise. Both probability density functions on the left are bell curves with symmetric tails that run out to plus and minus infinity. Both bell curves produce white noise centered at zero. Note the scale difference between the two types of white noise: Gaussian noise spikes range from 5 to –5 while Cauchy noise spikes range from –400 to 400. The slightly thicker tails in the Cauchy bell curve produce occasional gigantic noise spikes.

Cauchy noise is *impulsive*.

The Cauchy noise in figure 4.2 contains powerful noise spikes that can be more than a hundred times the magnitude of the humbler Gaussian noise spikes. The thicker tails on the Cauchy bell curve mean that extreme events have more probability of occurring than they do with the thinner-tailed Gaussian bell curve. The scale differs for the Cauchy noise plot to account for occasional noise spikes of large magnitude of at least −400 or 400 units. The plot of Gaussian white noise would look like a straight line if we overlaid it on the center of the plot of Cauchy noise. IQs do tend to obey the thinner-tailed Gaussian curve. We would sometimes find IQs of 5 and 500 if IQs obeyed Cauchy's bell curve.

There is another relationship between Cauchy and Gaussian noise. One of the ironies of mathematics is that the ratio of two Gaussian quantities gives a Cauchy quantity. So you get Cauchy noise if you divide one Gaussian white noise process by another.

The uniform noise in figure 4.1 also becomes Cauchy noise if one applies the trigonometric operation of the tangent. Consider a merry-go-round with all the children safely removed. Then mount a .22-caliber automatic rifle pointing straight out from the merry-go-round and pointing at a nearby wall. Now let the merry-go-round rotate and let the rifle fire "at random" according to the uniform probability rectangle in figure 4.1. Then the bullet marks on the wall will have an approximate Cauchy distribution.

Cauchy phenomena occur in their own right. They make good models of catastrophic events such as stock market crashes or corporate bankruptcies or biological malfunction at the heart or species level. They can also model various burst or impulsive noises in signal processing or radar communication systems. Statisticians also use a modified version of Cauchy bell curves in the so-called *t*-distributions. These are thick-tailed bell curves that we often use when the data sample is smaller than 30. Such bell curves start out as the thick-tailed Cauchy bell curve when there is only one sample. The tails get thinner as the number of random samples increases. Such bell curves eventually become Gauss's thin-tailed bell curve as the data pour in. Thus do fat tails reflect scientific skepticism about drawing conclusions from small sets of data.

There is a still deeper relationship between the Cauchy and Gaussian bell curves. Both belong to a special family of probability curves called *stable* distributions that the French mathematician Paul Levy introduced in 1925. Levy abstracted a simple but important property from Gaussian quantities—they are closed under addition. If you add two Gaussian noises then the result is still a Gaussian noise. This "stable" property is not true for most noise or probability types. It is true for Cauchy processes. Levy found precisely those random models for which it is true. Thus did Levy discover or invent stable probability distributions. We now call one of these impulsive noise processes a Levy motion in honor of his discovery and its generalization of classical Brownian motion.

Stable curves describe impulsive behavior. They use a numerical parameter called *alpha* that ranges from 0 to 2. Alpha controls the thickness of the bell curve (where here we assume that all the curves are symmetric although stable probability curves can be asymmetric). The tails grow thicker as alpha grows smaller. The Gaussian bell curve is the special case of the thinnest tails of all—when alpha equals 2. The Cauchy bell curve has much thicker tails and corresponds to the case of alpha equal to 1. These are the only known closed-form symmetric stable bell curves even though there are as many stable bell curves as there are real numbers. Complicated algorithms estimate the stable bell curves with fractional alphas as in figure 4.3:

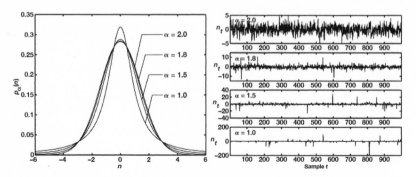

Figure 4.3: Four types of symmetric alpha-stable noise. The parameter α describes the thickness of a stable bell curve and ranges from 0 to 2. Tails grow thicker as α grows smaller. The white noise grows more impulsive as the tails grow thicker. The Gaussian bell curve ($\alpha=2$) has the thinnest tails of the four stable curves while the Cauchy bell curve ($\alpha=1$) has the thickest tails and thus the most impulsive noise.

The four stable bell curves in figure 4.3 show that white noise becomes more impulsive or more "bursty" as tail curves get thicker— and thus as the numerical parameter alpha falls in value. Scientists often call such noise *symmetric alpha-stable noise* to emphasize that a symmetric bell curve describes the probability of noise spikes (the alpha term relates to the so-called power law in the bell curve's characteristic function or Fourier transform). This turns out to be a special case of the general stable model.[7] A good analogy for stable noise is flashing blue lightning in a thunderstorm. Each lightning "bolt" is a random spike of sorts. The flashes would become more frequent and more powerful as the alpha value falls from 2 down to 0.

Far more complex noise types can arise from skewed asymmetric stable probability curves. Network engineers have shown that the distribution of bit packets on the Web or Ethernet likely obeys a skewed stable noise model with an alpha value of 1.7 or so.[8] That implies noise spikes for Web traffic even more violent than the ones in the second noise plot of figure 4.3 for the stable bell curve with alpha equal to 1.8.

This huge family of stable probability models enjoys two widespread properties in nature—and both challenge standard scientific descriptions of those processes.

The first property is a technicality but one rich with real-world consequences. It is that all stable probability models obey a *generalized* central limit theorem. The converse is also true: *Only* stable models obey such a theorem. So once again Gauss's important bell curve is not the unique bell curve. Nor does that bell curve hold some kind of monopoly on central tendency despite almost two centuries of practice to the contrary by the majority of scientists and engineers who have argued that adding up a lot of independent effects gives you Gaussian noise or some other Gaussian bell curve instead. The logic of that practice turns out to be a non sequitur in most cases. Adding up several independent effects does indeed lead to a bell curve. But it can be any one of the infinitely many stable bell curves. It takes data or experiment and not hand waving to decide the issue. This also helps explain the growing list of applications of stable models from interruptions in telephone lines and fluctuations in financial prices to fluctuations in underwater acoustics and the gravitational strength of planets with irregular mass distributions. There are now well over a thousand scientific and engineering articles that describe stable noise or stable bell-curve behavior in nature.

This bell-curve uncertainty gives new urgency to an old methodological question: Is an extreme data point a fluke or part of a legitimate pattern? A physician or economist could have dismissed such data as a mere "outlier" or the result of experimental error in the old days when Gauss's bell curve held a de facto monopoly on descriptions of bell-curve phenomena. The usual rule of thumb was and often still is that one can safely dismiss data that lies at least three standard deviations away from the mean or center point. The spread of stable techniques undermines such easy conclusions. The extreme data might well be part of a stable pattern with an alpha value of 1.9 or 1.5 or some other number. Again neither logic nor guesswork can decide the matter. Only direct measurement can tell. And there are many statistical tests that can render a verdict. Failure to check the stable status in some cases might even amount to professional negligence.

The second property is another technicality that has profound consequences for the sciences: Dispersion differs from variance.

Dispersion measures how data distribute or scatter about some central point. Dispersion in this sense is the footprint of randomness or uncertainty. Statisticians go to great lengths to estimate this crucial quantity because it can make or break the always imperfect fit between the pristine math of the model and the sloppy or fuzzy data of nature. Yet the common use of the term "standard deviation" shows that the assumption that dispersion equals variance has deep roots in our scientific worldview. This extends all the way to quantum talk of the uncertainty principle since such relations involve multiplying two standard deviations.[9]

The stable models challenge this assumption because stable probability curves have *infinite variance* and yet have finite dispersion. Many scientists dismiss infinite variance as physically impossible because they believe infinite variance requires infinite energy in the same way that pure white noise requires infinite energy. Or they dismiss it because they think that it implies so much uncertainty that such models could convey no information as in classical interpretations of quantum uncertainty.

Both notions are wrong in general. Stable curves behave differently than the finite-variance curves of most science textbooks where Gauss's bell curve is usually the only bell curve—and everything looks like a nail to a hammer. The misconception stems from extending Gauss's bell curve out wider and wider until it becomes nearly flat. This again is the classical interpretation of quantum uncertainty. If we have precise information about the position of a particle then we have imprecise information about its momentum or velocity. This means that if we have a narrow Gaussian bell curve for position then we have a fatter Gaussian bell curve for momentum or velocity. The velocity bell curve spreads out so wide that it becomes flat if the position bell curve collapses into a spike of total certainty. Then any velocity value is as likely to occur as any other.

The flaw in the classical thinking is the assumption that variance equals dispersion. Variance tends to exaggerate outlying data because it squares the distance between the data and their mean. This mathematical artifact gives too much weight to rotten apples. It can also result in an infinite value in the face of impulsive data or noise. Figure 4.4 shows just how wrong the classical thinking can be. It shows a stable bell curve (with alpha value equal to 1.8) that has infinite variance but that has three different types of finite dispersion that lead to three distinguishable types of white noise. It also turns out that some covarying stable noise variables can obey theorems that produce quantumlike "uncertainty relations" between the variables even though the noise variables have infinite variance.

Yet dispersion remains an elusive concept. It refers to the width of a probability bell curve in the special but important case of a bell curve. But most probability curves don't have a bell shape. And its relation to a bell curve's width is not exact in general. We know in general only that the dispersion increases as the bell gets wider. A single number controls the dispersion for stable bell curves and indeed for all stable probability curves—but not all bell curves are stable curves.

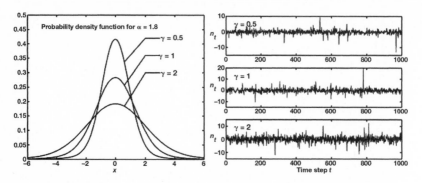

Figure 4.4: Dispersion differs from variance. The left figure shows three stable bell curves with tail thickness or alpha value of 1.8. The bell curves differ in their dispersion (gamma) values. The right figure shows the corresponding impulsive white-noise sequences. The widest bell curve has the most dispersion about its center and gives rise to the most energetic white noise.

4.4. CHAOS AND FUZZ CAN PRODUCE WHITE NOISE

White noise can also be chaotic as well as impulsive and the two can combine in ways that we have not imagined.

"Chaos" refers to systems that are very sensitive to small changes in their inputs. A minuscule change in a chaotic communication system can flip a 0 to a 1 or vice versa. This is the so-called *butterfly effect*: Small changes in the input of a chaotic system can produce large changes in the output. Suppose a butterfly flaps its wings in a slightly different way. That can change its flight path. The change in flight path can in time change how a swarm of butterflies migrates. That can change how a new tree fungus spreads throughout the world. That can change the economic fate of a developing country. That can lead to further environmental change or even a pattern of war among rival nations. The butterfly effect updates the old adage that one side can lose a war because a horseman never delivered a message because the horse lost a shoe for want of a nail.

Chaos can leave statistical footprints that look like noise. This can arise from simple systems that are deterministic and not random. Figure 4.5 gives an example. A simple recursive equation called the logistic equation gives rise to time-series values between $-\frac{1}{2}$ and $\frac{1}{2}$ that look "random" even though the equation fully describes how the oscillations evolve in time. The caption states the recursive equation that anyone can use to create their own chaotic time series starting with any initial value in the appropriate interval. This chaos plot also has a frequency spectrum (not shown) and it turns out to be nearly a flat line. The spectrum jitters about a flat line and so on average it looks flat across the given range of frequencies. That is the footprint of white noise.

The surprising mathematical fact is that most systems are chaotic. Change the starting value ever so slightly and soon the system wanders off on a new chaotic path no matter how close the starting point of the new path was to the starting point of the old path. Mathematicians call this sensitivity to initial conditions but many scientists just call it the butterfly effect. And what holds in

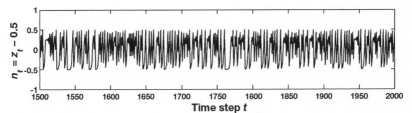

Figure 4.5: Chaos as whitelike noise. The shifted output of a "chaotic" logistic map appears as a noisy time series that bounces around randomly within the interval $(-5, 5)$. This noise plot has a frequency spectrum (not shown) that is nearly flat and thus nearly indistinguishable from that of white noise. Yet the noise n is *deterministic* because it arises from the deterministic logistic equation $z(t+1) = 4z(t)(1-z(t))$ where at each discrete time t the value lies in the interval $(-1, 1)$. The noise plot shifts the value by one-half to give the noise value n as $n(t)-\frac{1}{2}$.

math seems to hold in the real world—more and more systems appear to be chaotic.

Scientists and engineers have only begun to explore chaotic systems. Earlier scientists likely ignored chaotic data as due to error or "random chance" rather than as genuine patterns of nature. The spread of chaos in modern science creates a methodological problem similar to the one that impulsive noise creates for the analysis of extreme or outlying data: Is the data a chaotic pattern or just random noise? Neuroscientist Walter Freeman of the University of California at Berkeley has argued that our noses work with smell or olfactory patterns that are really chaos patterns. Other scientists see the same data as little more than random noise—and the debate continues.[10]

There is further complexity here. Fuzzy systems can produce signals that look like random noise or like chaos noise.

Fuzzy systems store commonsense rules of the form "If the air is cool then turn the air-conditioner down a little" and then make simple inferences by partially firing many of the rules and combining the results. They are fuzzy or vague because real-world concepts such as cool air do not have a bright black-and-white border between cool and noncool. Most borders are gradual and involve shades of gray.

A formal result called the fuzzy approximation theorem and its brethren guarantee that a fuzzy system can model any system to any given level of accuracy.[11] This fuzzy approximation works well for many small systems. That helps explain the thousands of gadgets from camcorders to microwave ovens that use fuzzy logic in their control chips. The same property lets fuzzy systems model random noise or chaos. A small number of well-chosen rules can create random-looking noise sequences for use in wireless communications. The U.S. Patent and Trademark Office issued a patent (patent number 5,539,769) to me and a former Ph.D. student for just such a fuzzy-noise system for use in the kind of spread-spectrum wireless communications that we discuss in the next chapter.

Fuzzy approximation systems are powerful because they use expert knowledge or real data to form rules and grow knowledge bases. But this also limits them. We often lack enough good experts to give us good rules or we lack enough good data to let us learn good rules through adaptive algorithms.

Fuzzy approximation also may not work well or work at all for large-scale systems because it can require too many rules. The number of rules tends to grow exponentially with the number of variables in the system. Such exponential rule explosion is an example of the so-called "curse of dimensionality" that bedevils all systems in some manner as they try to add more components to produce a more accurate model of the world. Physicists call a different version of the same curse the many-body problem. It explains why so far we can only fully solve the Schrödinger wave equation for the hydrogen atom or for "hydrogenlike" atoms or ions that have only one orbiting electron no matter how many protons and neutrons lie in the nucleus and not for the other elements in the periodic table.[12] Thus many properties of the periodic table reflect only approximations of the complex interactions of electrons that determine real chemical properties.

The curse of dimensionality also explains why we can't yet work out the exact path of a nearby asteroid if we take into account the full gravitational effects of the sun and the planets and the moon. The math of rocket science works best for small-scale systems—at least when we use brains and not computers to come up with the equations and then to manipulate those equations.

4.5. REAL NOISE IS COLORED NOISE BECAUSE ITS FREQUENCY SPECTRUM IS NOT FLAT

Real noise cannot be white because again pure white noise requires infinite energy. That means that real noise has a spectrum that is not a perfect flat line across all possible frequencies. The spectrum of real noise has ripples and curves because real noise spikes do not invoke all frequencies with equal energy.

Colored noise is any noise that is not white noise. That means colored noise has a nonflat spectrum at least over some large range of frequencies. That in turn means that the noise spikes correlate with one another if only slightly. This leads to jagged but still random patterns of noise spikes. Technically all real noise is colored to some degree since there are no ideal flat lines across frequencies or anywhere else in the real world.

Physicists have developed a color-coded scheme for colored noise based on the exponent of the inverse of the frequency or f. (This is another example of a so-called *power law* in statistics because the fixed exponent states the power to which the equation raises the frequency.) This color scheme admits variations but is about as close as a color-noise standard as we have in the sciences. The scheme says that the noise is white if the noise spectrum does not depend explicitly on the frequency f. That corresponds to the case of $1/f$ raised to the power zero because the zeroth power gives the constant value of unity: $1/f^0 = 1$. Pink noise falls off or decreases with the inverse of the frequency. So pink noise has a spectrum that falls off with the first power of the frequency or $1/f$.

Other noise colors have frequency spectra that fall off faster than pink noise. Brown noise has a spectrum that falls off with the inverse square $1/f^2$ and produces a spectrum that looks like Brownian motion. Black noise has a spectrum that falls off faster than

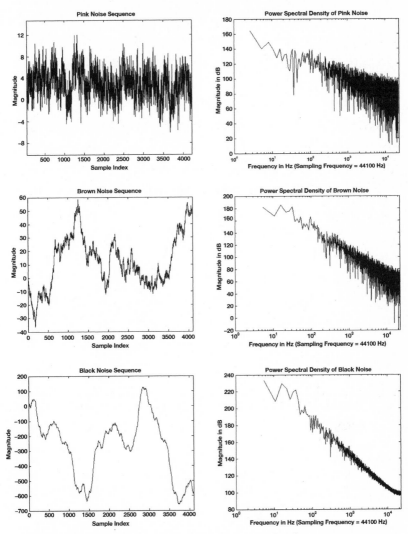

Figure 4.6: Colored noises in time and their power-law frequency spectra. Each figure on the left shows "colored" noise spikes in time. Each figure on the right shows the corresponding frequency spectrum as a function of the frequency f. The first pair shows pink noise or "flicker" noise with $1/f$ spectrum. The second pair shows brown noise with $1/f^2$ spectrum. The noise spikes of brown noise wander in a Brownian motion and often arise from a Brownian diffusion process. The third pair shows black noise with the steeper $1/f^3$ spectrum (note the scale change). Higher frequencies of real phenomena often fall off steeply as blacklike noise.

brown noise such as with the inverse cube $1/f^3$. Such fast falloff in frequency means that the total energy can be finite as we in fact observe in nature. Figure 4.6 shows both noise spikes and the frequency spectra of these power-law colored noises.

Pink noise is noise that the human ear hears as white noise. This has to do with the logarithmic nature of hearing (equal energy in octaves) and reflects our use of decibels to measure sound intensity. Musicians have experimented with pink-noise compositions because of its "self-similar" or fractal structure that arises from its simple power-law structure. Light with a $1/f$ spectrum need not appear pink but the name has stuck. Circuit engineers sometimes call this *flicker noise* because it can make the light-emitting filament in an old-fashioned vacuum tube flicker like a candle in the wind. Engineers sometimes use the name *popcorn noise* to describe pink or flicker noise in some computer chips.

Pink noise appears to be ubiquitous in nature—or so hundreds of journal articles have argued. Scientist have found various types of empirical and theoretical evidence for pink noise in a wide range of physical devices and natural phenomena. Electrical circuits and semiconductors emit pink noise over many decades or octaves of frequency. More speculative examples include the power spectra of extinction in the fossil record and the power spectra of tropical convective currents in the ocean. Pink noise in fossil data suggests long-range correlations among different extinctions over many millions of years despite a lack of any convincing mechanisms that could produce such correlated behavior. Earthquakes and avalanches also appear to occur in pink-noise fashion for at least part of their range. These last two examples of pink or pinkish noise are part of the more general phenomenon called *crackle noise*:

> Many systems crackle. When pushed slowly they respond with discrete events of a variety of sizes. The

Earth responds with violent and intermittent earthquakes as two tectonic plates rub past one another. A
piece of paper (or a candy wrapper in a cinema) emits
intermittent sharp noises as it is slowly crumpled or
rumpled. A magnetic material in a changing external
field magnetizes in a series of jumps.[13]

The vast scientific literature on pink noise reflects the substantial
controversy over whether or to what extent these "crackling" systems and other systems exhibit pink-noise behavior. The main exception is the consensus that electrical circuits exhibit flicker or pink
noise. There is even less agreement over the underlying causes of
such simple power-law behavior if it in fact occurs in nature.[14]

Brown noise sounds softer or less harsh than white noise or pink
noise. It too appears to be common in nature. Brown noise tends to
arise from the Brownian motion involved in many physical and
chemical diffusion processes. Temperature fluctuations in a given
city can look like brown noise over time. The time-series plots of
brown noise themselves look like "random walk" sample paths from
a randomly jittering Brownian motion as in the second time-series
plot in figure 4.6. Many pink-noise systems shade off into brown
noise at high frequencies.

The same holds for black noise. The spectra of physical systems
often decay faster than brown noise for high and very high frequencies. There is some evidence that river height fluctuations resemble
black noise but overall the evidence for this noise type is not nearly
so strong as for pink noise or brown noise.[15] Another problem is that
computing power spectra involves using any one of several fairly sophisticated algorithms from the field of signal processing. Each algorithm involves many assumptions and cutoff values and a great deal
of number crunching. So what appears to be black noise may simply
be a mathematical artifact of one of these algorithms.

A real-time signal tends to have several "colors" in its spectrum. It may have an approximately flat or white part followed by a pink-like fall in frequency that ends in a steep falloff akin to brown or even black noise. And again this noise may be impulsive or arise from underlying chaotic dynamics or a fuzzy structure.

We turn now to what is arguably the most ubiquitous whitelike "real" noise in nature—thermal noise.

4.6. THERMAL NOISE FILLS THE UNIVERSE

Thermal noise is a common and often vexing example of Gaussian white noise—of approximately Gaussian and approximately white noise. All heat sources produce thermal noise because of molecular collisions and vibrations. The intensity of thermal noise increases with the absolute temperature of a substance. Thermal noise also increases with the bandwidth or the swath of frequencies that a system uses to transmit bits or other message elements. So communication engineers always face more thermal noise when they use a wider bandwidth to send satellite signals or other signals.

Thermal noise is in the nature of things because it is directly proportional to temperature and because things made of molecules have temperature. Temperature itself is a statistical fiction or artifact. It is a number that reflects the *average* velocity of a large number of molecules. Raising the temperature of a substance means that its molecules move faster on average. The kinetic theory of gases further tells us that the temperature of a gas depends directly on the average kinetic energy of the moving and colliding gas molecules.[16] These independent collisions and vibrations produce thermal noise and invoke the central limit theorem to give thermal noise its Gaussian (or its thick-tailed stable) white structure.

Thermal noise is not truly white because thermal noise favors different types of electromagnetic radiation at different temperatures even though it produces some radiation of all wavelengths. Thermal noise for very cold sources near absolute zero tends to be mainly microwave radiation. There would still technically be some thermal noise at absolute zero (about −273.15°C or −459.67°F) even if quantum uncertainty did not forbid such a certain outcome. Heat or temperature energizes the collisions of electrons in a circuit or the collisions of molecules in other substances or gases.

To be is to emit thermal noise.

Our warm bodies produce faint thermal noise. Rattlesnakes can easily detect the infrared portions of it. The earth oozes thermal noise because of its hot interior and because its surface and atmosphere absorb about 70% of the radiation that arrives from the sun. This thermal noise lowers the signal-to-noise ratio of radio systems and some other types of communication. The sun is a powerful source of thermal noise because its superhot core burns hydrogen into helium ash in a nearly endless sequence of controlled thermonuclear explosions. The burning in the core will end in about 5 billion years when the sun expands as a red giant and engulfs the earth while it warms up Pluto. This nuclear fusion in the core takes place at pressures billions of times greater than our sea-level air pressure and at temperatures millions of degrees hotter than room temperature.

The distant stars produce enough collective thermal noise or so-called *sky noise* to interfere with antennas that point across the atmosphere and to interfere to a lesser extent with those that point straight up.[17] The antenna itself produces thermal noise that may interfere with sensitive astronomical measurements.

Radar systems suffer interference from *atmospheric noise* that arises from the 100 or so lightning flashes per second in the worldwide atmosphere. A lightning flash releases a massive electrical discharge and superheats an air column to about 50,000 degrees Fahrenheit or many times hotter than the surface of the sun. This rapid heating produces the pressure shockwave that we hear as a thunderclap. Such lightning-based radio-frequency noise of up to about 50 megahertz can interfere with radar systems.

Radar itself means "radio-wave detection and ranging." A radar system measures echoes or reflections from an expanding area of radio waves. The expanding area is part of a sphere and thus results an inverse-square law: Signal strength decreases with the square of the distance traveled. So the basic radar equation states that the strength of a received radar signal that bounces back from an object decreases with the *fourth* power of the distance to that object. This is rapid falloff in signal strength with distance. Radar interference can also arise from moisture or from other radar signals. Even insects can reflect a radar echo and count as part of the so-called *clutter noise*.

Albert Einstein laid the groundwork for the modern understanding of thermal noise when he published his famous 1905 paper on Brownian motion. Brownian motion gets its name from what English botanist Robert Brown saw under his microscope in 1827: Pollen grains in water jitter about at random. Suspicion grew over the decades that such Brownian motion resulted from water or gas molecules randomly colliding with pollen grains or other small particles much as if thousands of drunken athletes kicked and punched a large beach ball in a soccer field.

Einstein applied physical reasoning to conclude that such Brownian motion acted as a diffusion process that depends on temperature and friction. This means that Einstein showed that Brownian motion obeys the heat equation of classical physics. Einstein's other 1905 work on special relativity often overshadows this stunning finding. Still his reduction of Brownian motion to the heat equation is one of the great achievements of the atomic theory and indeed one of the great conceptual breakthroughs of all time.

The heat equation explains the temperature gradients we see on a weather map. The heat equation itself is a special case of the ubiquitous reaction-diffusion equation that describes diverse phenomena in all the sciences—from the fluctuation of quantum matter waves in space (the Schrödinger wave equation) to the spread of insects or fire in a forest to the spread of a rumor or a gene for green-eyedness in a human population. It states that how a quantity changes in time depends on how it concentrates or spreads out in space. Physicists have generalized the heat equation to include frictional forces and restoring forces that oppose the random Brownian disturbance. These and many other generalizations of Brownian motion still reduce to Einstein's probabilistic heat equation when they ignore these other forces.

Einstein proved another remarkable mathematical fact about Brownian motion: There is a probability curve that satisfies the heat or diffusion equation and it too is Gauss's bell curve.[18] Mathematicians much later proved that only Gauss's bell curve satisfies the heat equation and thus the solution is unique. Einstein's result explains why Brownian motions have a Gaussian probability structure. That in turn explains why the velocity or instantaneous rate of change of a Brownian motion is Gaussian white noise such as the first noise panel shows in the simulation approximation in figure 4.1.

Circuit analysis brought the next major advance in understanding thermal noise.

Thermal noise remains a practical and research problem for engineers who try to pack ever smaller electrical circuits on a computer chip. Inherent thermal noise can destabilize circuits by randomly flipping bits from 0 to 1 or from 1 to 0. Many engineers have speculated that thermal noise and related quantum noise will slow or even stop Moore's law—the empirical trend that the density of electrical circuits on a chip doubles every two years or so. But Moore's law celebrated its fortieth anniversary on 19 April 2005 with no end yet in sight. Chip makers now produce many more logic circuits or transistors each year than there are grains of rice.

Engineers long ago observed and deduced that the Brownian collisions and random vibrations of electrons in conductors produced white or near-white noise. This thermal noise grows with temperature or heat energy and does not require that current flow through a conductor. White noise from so-called *shot noise* does require current flow. Such shot noise stems from the independent random quantum effects of electrons or other energetic particles that cause tiny current fluctuations. Amplified shot noise can sound like several BBs or shotgun pellets falling on a hard tile surface. Scientists and engineers sometimes refer to shot noise as *Schottky noise* in honor of the German scientist who described the effect in a 1918 paper in the *Annals of Physics*. They later observed white thermal and shot noise for optical systems and more recently for some nanosystems. The term *quantum noise* sometimes refers to either shot noise or thermal noise or any other energetic fluctuation at the quantum level.

Engineer John Bertrand Johnson first carefully measured thermal noise in electrical circuits while he worked at Bell Laboratories in the 1920s. He reported his experimental findings in 1928 in the paper "Thermal Agitation of Electricity in Conductors" in *Physical Review*. His Bell Labs colleague Henry Nyquist found a mathematical explanation for the thermal noise that Johnson had described. Nyquist published his results in the same issue of *Physical Review*. We still refer to his theoretical noise finding as Nyquist's theorem.[19]

The result is that engineers and physicists sometimes call thermal noise Johnson noise or Nyquist noise or even Johnson-Nyquist noise. We also continue the working fiction of modern science and engineering and say that such thermal noise is white because its frequency spectrum is approximately flat over so much of the relevant frequency spectrum (almost up to a terahertz). Nyquist's theorem actually says that part of the spectrum of thermal noise will be sloped and not flat over a fair portion of the frequency spectrum. Again it is approximately flat over most frequencies of interest in a circuit that operates at or near room temperature. So even in theory white thermal noise is not so white. The endnotes give the technical details of this leading example of "real" white noise.

Our brains also emit thermal noise and they do so in at least two ways.

The first way is simply by being warm and weighing about three pounds or about 1.5 kilograms. Brains gorge on blood sugar or glucose and can quickly malfunction or die without it. The brain consumes about 20 watts of power each day. Some of this power dissipates away as waste thermal noise. This brain chain from sugar to noise reminds us that we are ultimately sugar parasites: Plants make the carbohydrate glucose $(C_6H_{12}O_6)$ and breathable molecular oxygen (O_2) through photosynthesis when sunlight energizes a chemical reaction based on carbon dioxide and water:

$$6CO_2 + 6H_2O \rightarrow C_6H_{12}O_6 + 6O_2$$

Then we eat the plants for their sugar or we eat the creatures such as cows that eat the plants for their sugar. Either way much of it ends up as sugar fuel for a warm and wet computer that leaks thermal noise in all directions. And it all starts with the visual white noise we call daylight.

Brains also emit thermal noise as part of their wet neural-network circuitry. The brain contains about 100 billion neurons. Each neuron is an electrochemical switch of sorts. And each neuron connects on average to about 10,000 other neurons in a massive neural network. The connections act like wires or rather leaky electrical cables and make up a great deal of the wet mass of the brain. Thermal noise pervades these tiny electrical circuits because of current resistance in each neuron's cell membrane and other forms of current resistance. This brain noise turns out to be small compared with noise from synapses and from those random openings and closings of potassium-ion and sodium-ion channels and from other ionic channels in each neuron.[20] Vast numbers of tiny ionic pumps underlie the generation and transmission of each neuron's electrical impulses. These ionic pumps burn up as much as half of the brain's glucose in the form of the energy molecule ATP (adenosine triphosphate). A further noise source comes from the extensive unwanted crosstalk from other flickering neurons. Still the presence of thermal noise does somewhat diminish each neuron's and each neural net's ability to process signals. The first and last chapters discuss the important exception called *stochastic resonance* where sometimes small amounts of such noise can help neural computation.

The ultimate thermal noise comes from the big bang.

The universe began in a superhot explosion about 14 billion years ago. It has radiated thermal noise ever since. The thermal noise has lessened as the universe has expanded and cooled because thermal noise depends directly on temperature. This cool and low-energy thermal noise exists in the form of faint electromagnetic radiation. So the radiation has long wavelengths because again basic quantum mechanics requires that high-energy radiation have short wavelengths and that low-energy radiation have long wavelengths. Physical theory and Wien's law predict that such residual radiation (with wavelength about 1.89 millimeters) from the big bang should occur as microwaves that radiate uniformly in all directions. Physicist George Gamow first predicted the existence of such a cosmic microwave background in 1948.

The evidence confirms the microwave prediction and much of the uniform-direction prediction. This led to a 1978 Nobel Prize in Physics for Bell Labs scientists Arno Penzias and Robert Wilson because they inadvertently discovered the faint 2.725 degrees kelvin microwave (159 gigahertz) radiation when working with a low-noise microwave receiver. NASA launched the Cosmic Background Emission (COBE) satellite in 1988 and further confirmed the microwave noise and its general uniform or "isotropic" distribution. It also found clumps of nonuniformities or "anisotropies" in the background radiation.

NASA next launched the Wilkinson Microwave Anisotropy Probe (WMAP) in 2001 to further study this ancient cosmic noise and its intriguing nonuniformities. Astrophysicists have proposed that these nonuniformities in the microwave spectrum may have arisen from any of several competing physical hypotheses about the clumpy formation of matter and galaxies in the universe. WMAP further confirmed the basic microwave prediction of the big bang hypothesis. The European Space Agency plans to launch the Planck Surveyor satellite in 2007 to more closely study the cosmic microwave spectrum.

The WMAP satellite has squeezed a great deal of information out of the cosmic microwave noise. Its data implies that the universe is 13.7 billion years old within about 1% margin of error. The first stars started burning about 200 million years after the big bang exploded from a superhot and superdense point smaller than a quark. WMAP also found that only about 4% of the matter in the universe consists of ordinary atoms and molecules. The other 96% consists of mysterious dark energy and so-called cold dark matter. That still gives enough matter for us to conclude that the universe is flat and not curved. So the universe will likely expand outward forever and grow only colder as it dies in a "cold death."

4.7. EVEN BLACK HOLES EMIT NOISE—AND DIE

Black holes may be our last hope for a long-term energy source. But even they do not last forever because they leak heat as thermal noise and so they too eventually die.

Black holes have such intense gravity that light cannot escape from them. So they should appear black. Astronomers believe that a large black hole resides at the center of our Milky Way galaxy and may tear apart and consume nearby stars. The earth would form a black hole if we somehow compressed it down to the size of about a marble. That would cross the critical limit where the dense object's gravity would in effect turn in on itself and suck all its matter down to a point or "singularity." A marble-size black region or event horizon would surround the infinitesimal singularity in the space-time continuum. The sun is not massive enough to become a black hole when it burns up the hydrogen in its core in about five billion years. It will instead expand into a red giant and then cool off and die quietly as a white dwarf. The sun would become a black hole if we could compress it to a dense ball with a radius of about one kilometer.

Chapter 1 pointed out that black holes have an entropy or uncertainty. This makes intuitive sense since black holes represent pure physical uncertainty about the structure of the matter that fell into them. Physicists refer to this property by saying that "black holes have no hair" even though at least one model says that a black hole has some hair memory in the form of strings.[21] A black hole has a corresponding amount of information that is proportional to the square of its mass. This leads to the "it from bit" thesis of physics: The world is made of bits of information. A giant black hole that swallowed the whole universe would contain about 10^{120} bits of information. This huge cosmic bit count dwarfs Claude Shannon's 1950 estimate of 10^{43} possible chess games that consist of only 40 moves. But the size of the cosmic bit count is still likely less than the as yet unknown exact number of all possible chess games.

Physicist Stephen Hawking showed in 1974 that black holes should give off thermal radiation and thus should emit electromagnetic noise. This led to Hawking's colloquial phrasing that "black holes ain't so black" in his popular book *A Brief History of Time*.

Hawking radiation occurs when a black hole eats a fraternal

twin of a particle such as a photon of light. It pays for this meal out of its own mass. Pairs of virtual particles flash quickly into and out of existence from the not so empty vacuum of space. The pair could be an electron and a positron or any two other dual particles. But Heisenberg's uncertainty principle implies that there can be no true vacuum because that would entail a certain or nonprobabilistic relationship between energy and time. This uncertainty relation further implies that an emitted photon of blue light will last only about a billionth of a billionth of a second. Such particle pairs regularly flash into existence just outside the black hole. Sometimes the black hole sucks one of the particles into its singular belly. Then the surviving particle appears as Hawking radiation and has the form of quantum blackbody radiation. The fleeing particle takes energy from the black hole and thereby reduces the hole's mass.

Smaller black holes produce surviving particles that have more energy than do the surviving particles that larger black holes produce. Tiny black holes can in theory give off substantial noise because the temperature of Hawking radiation varies inversely with its mass.[22] So massive black holes at the center of galaxies emit almost no such radiation while tiny black holes can emit tremendous amounts of radiation. The power or wattage of the emission also grows inversely not with the black hole's mass but with its square. So the smallest black holes shine the brightest and burn the hottest.

Tiny primordial black holes may have formed just after the big bang exploded into existence. These primordial black holes should emit detectable light in the universe because the universe is only about 14 billion years old and because primordial black holes should live for only a few billion years. Indeed tiny primordial black holes should explode as their mass rapidly burns up in Hawking radiation of high wattage. Sensitive satellites should be able to detect gamma rays and perhaps other radiation signals from such tiny explosions. So far they have not.

Large black holes that swallow up stars or whole galaxies of stars would take eons to evaporate away through Hawking radiation. Evaporation time increases with the cube of the black hole's mass. Giant holes could take on the order of a googol (10^{100}) of years to evaporate. But in time even these great potential stores of energy will evaporate. So advanced civilizations could not extract energy from them forever even if their residents could take ever longer naps as the universe thinned out. They must succumb in time to the cold death of endless expansion as even the giant black holes run out of energy. By then all other matter will have radiated itself away in thermal noise all the way down to a cold fraction just above absolute zero. It will be the Big Chill—*forever*.

All that will remain of prior existence will be a very faint noise.

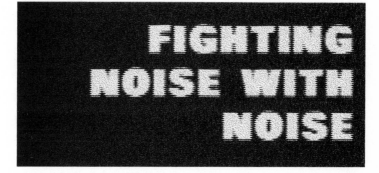

CHAPTER 5

FIGHTING NOISE WITH NOISE

Annoying noise in the passenger cabins of propeller aircraft, the rumble in air-conditioning systems, and the sounds disrupting headset communications are being reduced these days by active noise control, thanks to advances in digital signal processing. The technique relies on the principle of destructive interference between two sound fields.

> –Stephen J. Elliot
> "Down with Noise"
> *IEEE Spectrum,* June 1999

The total amount of information which may be transmitted is proportional to the product of the frequency range which is transmitted and the time which is available for the transmission.

> –Ralph V. L. Hartley
> "Transmission of Information"
> *Bell System Technical Journal,* vol. 7, 1928

We can say that any function limited to the bandwidth W *and the time interval* T *can be specified by giving* 2TW *numbers.*

　　—Claude E. Shannon

　　"Communication in the Presence of Noise"

　　Proceedings of the Institute of Radio Engineers, vol. 37, 1949

Under certain conditions a continuous-time signal can be completely represented by and recoverable from knowledge of its values or samples *at points equally spaced in time. This somewhat surprising property follows from a basic result that is referred to as the* sampling *theorem.*

　　—Alan V. Oppenheim and Alan S. Willsky

　　Signals and Systems

Hitherto communication theory was based on two alternative methods of signal analysis. One is the description of the signal as a function of time. The other is Fourier [frequency] analysis. Both are idealizations, as the first method operates with sharply defined instants of time, the second with infinite wave-trains of rigorously defined frequencies. But our everyday experiences—especially our auditory sensations—insist on a description in terms of both *time and frequency.*

　　—Dennis Gabor

　　"Theory of Communication"

　　Journal of the Institute of Electrical Engineers, 1946

Any desired spectrum $P(\omega)$ *can be obtained by passing wideband resistance noise or "white" noise through a shaping filter whose gain characteristic is* $\sqrt{P(\omega)}$.

　　—Hendrik Wade Bode and Claude E. Shannon

　　"A Simplified Derivation of Linear Least Square

　　Smoothing and Prediction Theory"

　　Proceedings of the Institute of Radio Engineers, vol. 38, 1950

The essential and principal property of the adaptive system is its time-varying, self-adjusting performance. The need for such performance may readily be seen by realizing that if a designer develops a system of fixed design *which he or she considers optimal, the implications are that the designer has foreseen all possible input conditions, at least statistically, and knows what he or she would like the system to do under each of these conditions.*

—Bernard Widrow and Samuel D. Stearns
   Adaptive Signal Processing

The basic principle of echo cancellation is to synthesize a replica of the echo and subtract it from the returned signal.

—Simon Haykin
   Adaptive Filter Theory

Thermal noise in electronic circuits, which is usually a nuisance to be suppressed, becomes a resource to be exploited in random number generators. In one scheme the noise signal is measured at regular intervals defined by a sequence of clock pulses. If the voltage at the instant of a pulse is positive, a 1 is emitted, and otherwise a 0.

—Brian Hayes
   "Randomness as a Resource"
   American Scientist, vol. 89, July 2001

This invention relates broadly to secret communication systems involving the use of carrier waves of different frequencies, and is especially useful in the remote control of dirigible craft such as torpedoes. An object of the invention is to provide a method of secret communication which is relatively simple and reliable in operation, but at the same time is difficult to discover or decipher.

—United States patent number 2,292,387,
   "Secret Communication System"
   issued on 11 August 1942
   Coinventors: Hedy (Lamarr) Markey and George Antheil

> *Spread spectrum communication technology has been used in military com-*
> *munications for over half a century, primarily for two purposes: to cover the*
> *effects of strong intentional interference (jamming) and to hide the signal*
> *from the eavesdropper (covertness). Both goals can be achieved by spread-*
> *ing the signal's spectrum to make it virtually indistinguishable from back-*
> *ground noise.*
>
> —Andrew J. Viterbi
> *CDMA: Principles of Spread Spectrum Communication*

Much of the war on noise takes place in the modern field of signal processing. This field grew out of the study of analog electrical signals in the early twentieth century. Analog signal processing developed into DSP or digital signal processing as old and big analog computers gave way to smaller and more powerful digital computers and as analog storage systems gave way to digital storage systems. The compact disc is the exemplar of this digital triumph.

The DSP arsenal contains many weapons to cleanse a measured signal of noise contamination. This chapter looks at some of these weapons that use noise or noiselike structures to combat noise.

The first of these techniques allows us to reconstruct a properly sampled signal with an approximation of white noise. It also offers a practical way for the reader to see some of the important engineering tradeoffs that underlie modern communication systems—especially the fundamental tradeoff between precision in time versus precision in frequency. The second technique is noise shaping. It shows how we can in theory use white noise to create almost any kind of practical signal from a birdsong to someone's voice. The third technique is adaptive noise cancellation. It shows that the best way to get rid of a changing noise source is to build a good mathematical model of the ambient noise and then subtract it from the system. Adaptive noise cancellers can track and suppress changing noise signals while anti-noise physical structures must do their best to suppress only an average level of noise. The final technique is spread spectrum. It shows how we can hide signals in synthetic noise so that an eavesdropper hears or sees only faint white noise. Spread spectrum can also resist most attempts to jam a signal and lets several signals efficiently occupy the same wide swath of frequency bandwidth. Spread spectrum underlies much of modern wireless communications and it arguably began in a 1942 patent from the film actress Hedy Lamarr.

5.1. THE IDEAL LOW-PASS FILTER RESEMBLES WIDEBAND NOISE IN DIGITAL SAMPLING

The filter lies at the heart of digital signal processing. A filter lets some signals through and keeps out or filters out others. A coffee filter or strainer works on the sample principle. It lets the liquid and small coffee grounds through but keeps out the larger coffee grounds.

DSP focuses on "time signals" such as speech or music that vary with time. The filtering itself tends to take place in the frequency domain. The first chapter explained how a Fourier transform converts a time signal into its frequency components much as a prism splits white light into its component colors or frequencies. Engineers use Fourier transforms and many other transforms to convert time signals to frequency signals because often we can more easily process the signals in the frequency domain. Then a reverse or "inverse" transform converts the processed frequency signal back into a new time-varying signal for human consumption.

The simplest type of filter chops off or discards large portions of a signal's frequency spectrum. Real time-varying signals have energy distributed across wide portions of the frequency spectrum.

The DSP process begins by truncating or "preprocessing" these frequency signals to some manageable interval—and there begins the degradation in sound from a real signal to a digital signal.

The archetype filter is the ideal low-pass filter.

The ideal low-pass filter discards high frequencies in a speech or music signal and keeps lower frequencies. The ear tends to hear some of these higher frequencies as noise or not hear them at all. Yet filtering out too many high frequencies produces a strained and tinny sound as when you speak through a megaphone or through the cardboard tube inside a roll of paper towels. Phone systems discard frequencies above 4 kilohertz (4,000 cycles or oscillations per second) even though speech signals can contain a few high frequencies above 10 kilohertz. Music contains even more high frequencies. This low-pass truncation lets us speak to one another on phones but with nowhere near the fidelity of live face-to-face speech. Images define two-dimensional signals and also benefit from low-pass filtering to clean up pixel noise or to remove image splotches and other processing artifacts.

Low-pass filtering explains why FM radio sounds better than AM radio—AM uses less bandwidth than FM does. AM radio protocols limit sound signals to 10 kilohertz while FM radio uses almost twice as large a swath of frequencies. That means AM low-pass filtering chops off more of the high-frequency components of its signals. Then "modulation" shifts these narrow frequency bands to designated locations of the frequency spectrum where we can locate them on the radio dial. Modulation resembles the way different instruments in the orchestra can play the same melody at different registers that range from the low frequencies of the double bass up to the higher frequencies of the piccolo.

The ideal low-pass filter looks like a rectangle or frequency "brick wall" as in the five ideal low-pass filters in the left column of figure 5.1. A low-pass filter passes the frequencies inside the rectangle and

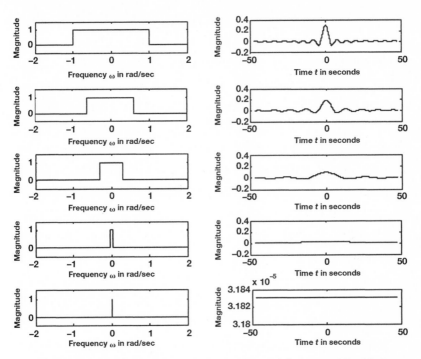

Figure 5.1: The tradeoff of resolution in time and frequency. The ideal low-pass filters in the left column define rectangles on the frequency axis. These "brick wall" filters pass low frequencies near zero and filter out or eliminate higher frequencies. The corresponding time versions of these ideal low-pass filters define the undulating sinc functions in the right column. The main lobe of the sinc wave spreads out as the frequency rectangle gets narrower and thus the time resolution degrades as the frequency resolution improves. Perfect frequency resolution gives a spike in frequency and a flat curve or no resolution in time as in the bottom row. Perfect resolution in time occurs when the sinc wave collapses to a spike and thus when the frequency rectangle expands out to infinity and approximates the flat spectrum of white noise.

filters out or excludes all other frequencies. It assigns the energy value of zero to all frequencies outside the rectangle. This simple filter is a special case of a *frequency selective* filter. A dual filter is the ideal high-pass filter that passes high frequencies and filters out low frequencies. The two filters differ in where the rectangle lies on the frequency axis.

Non-ideal low-pass filters can have almost any shape but most of their area will be over the "pass zone" of frequencies.

The ideal low-pass filter is ideal in at least one sense that white noise is ideal—we cannot physically realize it. This reflects one of nature's (or math's) harsh tradeoffs: A signal has a finite-length frequency spectrum only if it lasts infinitely long in time. So a finite spectrum implies infinite time and vice versa. The reverse also holds in the ideal world of mathematics: A signal is finite in time only if it has a frequency spectrum that is infinite in extent.

This tradeoff has a precise effect for a frequency rectangle such as an ideal low-pass filter. It implies that the time version of the low-pass filter defines a unique and symmetric undulating wave. The wave peaks at a particular value in time and then undulates away to the left and to the right at regular intervals. The right column of figure 5.1 shows portions of these so-called sinc functions in time.[1] The left and right undulations run all the way out to infinity on the horizontal axis. The undulations die down but the local peaks occur at regularly spaced intervals across the horizontal axis.

Related time effects occur for the hundreds of approximations to the low-pass frequency rectangle that signal processing engineers design. Changing the rectangle even slightly can produce complex oscillations in time. It takes sophisticated training to know which one of these approximate filters works best for a given real-world filtering problem. That is just one reason why graduating college seniors with a bachelor's degree in electrical engineering often command the largest entry-level salaries across all degrees.

The ideal low-pass filter reveals another deep and related tradeoff in the world of signals: We trade off precision in time for precision in frequency and vice versa. Perfect resolution in frequency implies no resolution in time and vice versa.

Notice that the rectangles in figure 5.1 get narrower as one moves down the left column. That means they pass fewer frequencies and thus the narrower rectangles give better frequency resolution. But this makes the corresponding sinc functions in figure 5.1 get wider. The frequency rectangle will eventually get so narrow that it collapses into a spike above just one frequency on the entire frequency axis. The result is an ideal filter for a single frequency and thus one that can perfectly resolve a selected frequency. But such ideal precision comes at the expense of making the sinc wave spread out as its own rectangle across the entire time axis. Then the flat sinc "wave" gives no time resolution at all as in the final graph in the lower right-hand corner. Note that the frequency axes in the figure include negative frequencies as well as positive or "real" frequencies. Such negative frequencies simplify the mathematics even though physical signals such as light and sound have only positive frequencies or cycles per second.

The reverse happens as the frequency rectangles in figure 5.1 get wider. Then the main lobe of the sinc functions get narrower and the undulations die down. This continues until the sinc collapses into a spike above just one time value (not shown). A spike in time perfectly resolves that instant of time. But then the frequency rectangle has grown to cover the entire frequency axis and so resolves no frequency at all. We saw in the last chapter that such a flat spectrum corresponds to ideal white noise. The same tradeoff occurs if we define a rectangle or "window" in the time domain because that produces an undulating sinc wave in the frequency domain. This opens the door to a variety of signal and noise processing techniques that lie beyond the scope of this discussion.

The tradeoff between resolution in time and resolution in frequency qualifies as a bona fide "uncertainty principle" of signal processing.[2] Decreasing a time signal's uncertainty implies increasing the uncertainty of the frequency version and vice versa. Versions of this uncertainty tradeoff appeared at Bell Laboratories and elsewhere around or before 1927 when physicist Werner Heisenberg independently published his famous quantum version of the uncertainty principle. Gauss's bell curve minimizes the overall uncertainty in all cases. The Oxford vision scientist John Daugman and his colleagues have produced statistical evidence that at least kitten brains process visual stimuli with cortical filters that resemble the mathematical filters called *Gabor logons* that minimize the time-frequency uncertainty.[3] This suggests that evolution long ago hit upon such optimal filters in the design of at least some neural mechanisms. The Hungarian electrical engineer Dennis Gabor introduced these uncertainty-minimizing "logons" in his famous 1946 paper "Theory of Communication" that reexamined the time-frequency uncertainty principle of signal processing. Gabor invented holography the next year and later won the 1971 Nobel Prize in Physics for that invention.

The ideal low-pass filter can reduce noise by chopping off the noisy portions of an observed spectrum. Noise usually has a wider frequency distribution than does a signal of interest such as music or human speech. The noise signal is "broadband" while the signal of interest is comparatively "narrowband." Again ideal white noise has a flat spectrum across the entire infinite extent of frequencies.

The irony is that the ideal low-pass filter can help signal processing by acting as a form of broadband noise. It does this indirectly as part of its role in the famous sampling theorem of DSP. This theorem dates back at least to the 1940s work of Claude Shannon at Bell Laboratories and likely much earlier given Shannon's own citations to prior work.[4] Engineers often call it the Shannon sampling theorem. More recently they have called it the Nyquist sampling theorem in honor of Bell Laboratories engineer and physicist Henry Nyquist whose noise theorem we discussed in the last chapter. Nyquist published an early version of the sampling theorem in a 1928 paper but he gave only a heuristic or partial proof. Shannon and others gave proper mathematical proofs of the theorem and its variants. This reflects the common logical ground between science and mathematics and even the law: It is not enough to get the right result—you also need the right reasons or justification for the result. You can yell "Heads" during a coin flip that turns up heads without having had any basis to claim that heads would occur. Stating an accurate conclusion differs from stating an argument or set of reasons that logically implies the conclusion.

The sampling theorem says that sometimes we can construct a continuous signal from discrete samples of the continuous signal. The theorem states that we can in theory *perfectly* reconstruct a continuous time signal from discrete samples or points in the signal if four technical and ideal conditions hold.

The sampling theorem reflects the artist's intuition that we can approximate the image of a face if we use enough small dots of color. It likewise reflects the working medical assumption that we can accurately measure a patient's smoothly varying blood pressure over an hour by taking measurement samples every five minutes or every minute or even every second. More discrete samples tend to give a better approximation of the face or the blood pressure. The sampling theorem states the remarkable fact that we get not just an approximation but the actual continuous signal itself from enough discrete samples if the four technical conditions hold. Much of the field of DSP deals with what happens when these ideal conditions fail to hold in practice—and they always fail to hold to some degree.

The four technical conditions show that there is no digital free lunch.

The first condition is that the signal must not have an infinite frequency spectrum. The signal must be "band limited." Again, real signals leave energy footprints over wide swaths of the frequency spectrum. So engineers have to make a judgment call and chop off the left and right tails of the signal's frequency distribution. This amounts to applying an ideal low-pass filter to the spectrum. It makes the information processing easier but it degrades the final quality of the reconstructed voice signal or music signal or TV image.

The second condition is that the number of samples is infinite but only countably infinite. This means that we could always count or enumerate all the samples up to any given sample. Such sample spikes look like an infinitely long comb with varying heights of comb teeth. The countable spikes in the two-dimensional case of image processing look like an infinite bed of nails.

Laymen may find it shocking or pointless to learn that there is more than one type of infinity. Indeed there are infinitely many types of infinity. There are as many types of infinity as there are counting numbers or integers. A famous and unresolved conjecture called Cantor's generalized continuum hypothesis says that there are only countably many infinities and no more—in particular not as many infinities as there are points on the real number line. Countable infinity is the simplest type of all and has the Hebrew-Latin name "aleph null" in the infinite hierarchy of infinities. It corresponds to the counting numbers whereas the continuous infinity involved in a continuous time signal corresponds to the entire real number line. The mathematical complexity of the sampling theorem lies in its relationship between these two types of infinity—in getting something continuous from something discrete. We can never achieve this ideal condition because in practice we have only a finite number of samples from any signal.

There is a second problem with the discrete samples that arises even if we could get a countably infinite set of them. The sampling theorem requires that the sample values *exactly* equal values of the sampled continuous signal. But we must "quantize" or round off real samples to some manageable number that we can represent with a string of 1s and 0s. This produces a quantization error or another form of noise in the process.

The third condition is the most famous condition for digital sampling. It states that the sampling rate or closeness of the samples must exceed twice the largest frequency value in the signal. This is the so-called *Nyquist frequency* or the Nyquist sampling rate. The Nyquist sampling rate of our 4,000 hertz phone lines is 8,000 hertz or 8,000 samples per second.

The Nyquist sampling rate is the key in practice. Sampling at a rate at least twice the highest frequency value ensures that each ideal low-pass filter rectangle in the frequency domain completely contains or passes all the frequency content of the original signal (which now has infinitely many replications on the entire frequency axis). Picture a rectangle as in figure 5.1 fitting around a tent or other shape of the signal's frequency spectrum. The infinitely many samples and sinc functions will always give back the original continuous time signal if the rectangles enclose the frequency tent or waveform. That occurs if the sampling rate is at least twice the highest frequency present in the signal. This could never happen if the signal did not have a finite-length frequency spectrum because then no finite-length rectangle could enclose the signal's frequency spectrum—just as a small cardboard box cannot cover all the golf balls that one can place on a golf green.

The Nyquist rate explains why music CDs record continuous music signals at a 44.1 kilohertz sampling rate or with 44,100 discrete samples per second for each stereo channel. Engineers chop off frequencies higher than 20 kilohertz because the ear tends not to hear them. Double that gives 40 kilohertz for a minimal Nyquist sampling rate. Add about 10% more samples to account for round-off or quantization effects and other design effects. The result comes to about 44.1 kilohertz. The CD encodes each discrete sample in a bit string of 16 0s and 1s. Tiny etched pits encode the bits in a spiral on the CD's plastic-coated aluminum surface. A 1 corresponds to the change from a pit to peak or vice versa while a 0 corresponds to no change from pit to pit or from peak to peak. An hour of music requires that the CD store over 5 billion bits in its miles-long spiral of etched pits. It also uses clever error-correcting (Reed-Solomon) coding to correct for random errors or missing data. So you can often scratch a CD with a nail and not hear the scratch when the music plays. The same nail scratch would degrade and maybe destroy an older analog vinyl LP album.

Sampling rates lower than the Nyquist rate produce a form of stroboscopic distortion or error called *aliasing*. A visual example occurs in some western movies when the wood-spoke wheels of a stagecoach or carriage appear to move forward and then backward as the stagecoach speeds up. The film consists of only 30 or so frames or samples per second. So aliasing occurs when an object moves faster than half this or 15 hertz. This occurs for large wagon wheels because their wooden spokes are both symmetrically spaced and indistinguishable from one another and because the wheels can rotate at high velocities. Recurring patterns of aliasing appear as the wagon wheels rotate faster. They appear at first to rotate normally in the forward direction because the photo sampling still obeys the Nyquist rate for the relatively slowly moving spokes. The spoked wheels rotate faster until they exceed the Nyquist rate and then aliasing makes them appear to rotate backwards. Next they will appear not to rotate even though the stagecoach or carriage moves faster than ever. The pattern of forward and backward and no rotation repeats as the wheels rotate faster still. So one way to tell if your brain is really a computer chip or if you are trapped inside a *Matrix*-like virtual world is to check rapidly rotating objects for aliasing effects.

The fourth and final condition also involves the noise connection. The condition states that we can reconstruct a continuously varying time signal such as a speech signal from infinitely many discrete samples if we *add* together infinitely many sinc functions of the type in the right column of figure 5.1. Each sample gets its own sinc function centered right on top of it. These sinc functions or wavelets *interpolate* between the discrete samples and give back the original continuous signal exactly. Real systems can add together only a finite number of such functions or approximations to such functions. This too further degrades system performance.

An alternative but more complicated system sums irregular *Daubechies wavelets* rather than the symmetrically undulating sinc functions. Engineers call such systems simply "wavelets." The Daubechies wavelets take their name from Princeton mathematician Ingrid Daubechies who discovered them in the late 1980s. Daubechies's work launched the modern field of wavelet analysis that seeks to jointly resolve signals in both time and frequency. The Federal Bureau of Investigation uses a discrete wavelet transform to efficiently compress and store the millions of digitized fingerprints in its databases.

The main noise connection is in the width of the frequency rectangle. It again has to be wide enough to pass the entire frequency spectrum of the original time signal. So it must be a broadband signal relative to the signal of interest. This is a fine point and one often lost in both the use and teaching of digital sampling. Still it shows an important but indirect way that noise assists modern DSP.

5.2. NOISE HELPS SHAPE THE SPECTRUM OF SIGNALS

Noise plays a more direct role in the shaping of frequency spectra. We can use white noise to create a signal with almost any frequency spectrum imaginable. Then the new frequency structure gives a new time-varying signal.

This simple but powerful fact goes under the names *noise shaping* or sometimes *spectral factorization*. It means that in theory we can take the white noise based on a computer's random number generator and use it to model the spectrum of a spoken word or the unique song of an eastern bluebird. Then a simple mathematical operation yields a synthesized spoken word or synthesized birdsong. A good enough DSP system could use white noise to create or "drive" synthetic speech samples that listeners likely could not distinguish from our own unique pronunciations. Some versions could also filter out interfering noise that would tend to corrupt or distort the speech spectrum.

Noise shaping uses the idea that sometimes you can figure out one part of a three-part system if you know the other two parts. This holds for the linear systems that underlie most DSP systems. The world itself may not be linear except to a first approximation. But engineers can often build devices that act linearly over wide ranges of inputs. More general nonlinear systems do not permit one to deduce one part from the other parts.

Here a linear system corresponds to a sheet of typing paper while a nonlinear system corresponds to a crumpled sheet. Yet even a severely crumpled sheet looks flat or linear if you zoom in close enough. Engineers and physicists use a similar mathematical trick to "linearize" many complex systems that might otherwise defy mathematical analysis or computer modeling. This linear approach to the world reached its zenith in the Shannon era of the mid-twentieth century before researchers had access to powerful and inexpensive computers and a wider range of mathematical tools and algorithms. The linear approach still remains the core around which researchers and practitioners experiment with more exotic techniques.

Classical systems theory divides a process or system into three parts. The first two parts are the input and output. Often we can directly observe these as when we put a dollar in a vending machine and get a can of cola as the output. The third part is the throughput or all that lies between the input and output. The throughput of real systems can have stunningly complex structure as between a car's accelerator and the spinning of the wheels or the thousands of biochemical processes that take place within our digestive tract.

The throughput of a linear system is a line of sorts. That means in the simplest case that the output is just a scaled or shifted version of the input. Doubling the input always doubles the output. The kinetic energy of a speeding car is not a linear system but a quadratic system. Doubling the car's speed or velocity quadruples its kinetic energy—and so high-speed head-on collisions are often fatal.

Linear DSP systems are slightly more complex. They have the extra property that their structure does not vary with time. That makes them "time invariant" as well as linear. This in turn implies that a linear filter has a so-called convolutional structure.

Convolution means that we reverse the order in which we multiply certain pairs of numbers.[5] This technical condition presents no problem for modern DSP chips because the required number crunching remains only the simple operations of multiplication and addition. The big payoff comes in the frequency domain after we pass the time signal through the prismlike structure of a Fourier transform. Then we can simply multiply frequency spectra. This happens in the sampling theorem when a low-pass filter rectangle chops off or filters out part of the frequency spectrum of a time signal. The chopping is in fact the multiplication of the two spectra and that arises from a formal convolution in the linear system. Engineers often call this the convolution theorem: Convolution in time gives multiplication in frequency.

Noise shaping exploits the convolutional structure of a linear DSP system by feeding white noise into the system as the input. White noise has a flat spectrum that looks like an infinitely long rectangle. The strength of the noise defines the height of the rectangle: Stronger noise gives a higher rectangle. The height is a single number that characterizes the entire frequency structure of the white noise. This gives an easy way to compute the frequency spectrum of the system output by multiplying the height by a form of the throughput.

The linear math implies that the noise height need only multiply the squared value of the throughput's frequency spectrum to give the output spectrum.[6] Engineers call the throughput frequency spectrum the "frequency response" of the system. Suppose the noise height is 4 and the value of the throughput spectrum or frequency response is 3 at some particular frequency. Then the corresponding value of the output frequency spectrum or the so-called power spectrum is 36 because $4 \times 3^2 = 4 \times 9 = 36$.

The opening epigraph from Bode and Shannon shows that engineers long ago saw that they could turn this around to shape or create just about any desired output frequency spectrum and the corresponding time-varying signal. The trick is to pick the throughput frequency values as the *square root* of the desired output frequency-spectrum values. Then the linear system will square this and then simply multiply the squared value by the noise height. The square-root filter "colors" the input white noise because it introduces correlations into the process. So some authors refer to such a filter as a coloring filter. They refer to the reverse or inverse filter as a whitening filter if it takes the output and gives back white noise. This square-root trick comes with an important technical proviso. The output power spectrum does not preserve phase information in the signal and such information can sometimes be crucial to the signal's qualitative or statistical properties. More advanced techniques allow engineers to maintain this phase information but at the cost of greatly increasing the computational complexity.

Consider again the previous example where the end result was an output frequency value of 36. Suppose now that we started with this number because for some reason of engineering design we want a system that produces a spectrum intensity value of 36 at that particular frequency. Then we work backwards to find the system or filter structure that gives that result when we feed white noise into the system. The square root of 36 is 6. So we want a system throughput such that at that frequency its spectrum has the value 6. That means we need white noise with a rectangle height of 1 because $1 \times 6 = 6$. Done.

The engineer can almost always do this if the power spectra satisfy a technical condition called "rationality" as most do in practice.

Speech synthesizers often use this noise-shaping technique and a small number of numerical values called "poles" to produce artificial speech sounds that match a given spectral pattern. A noise-shaping filter with just two poles can also almost perfectly reproduce the historical cyclic pattern of sunspot data where the sunspot cycle repeats approximately each 11 years in terms of the maximum number of sunspots[7] (even though the magnetic polarity reverses with each cycle to give its own 22-year cycle). The error or difference between the real sunspot data and the noise-shaped filter is slight and leaves spectral footprints that look like a low-intensity white noise.

Music CDs use a similar noise-shaping trick called "dithering" to make the round-off errors in the recorded samples sound like faint white noise. Rutgers electrical engineer Sophocles Orfanidis explains the process:

> In a digital audio recording and mixing system where all the digital processing is done with 20 bits, the resulting audio signal must be rounded eventually to 16 bits in order to place it on a CD. The rounding operation [quantization] can cause unwanted granulation distortions. Adding a dither signal helps remove such distortions and makes the quantization noise sound like steady background noise. Further noise shaping can make this white noise even more inaudible by concentrating it onto spectral bands where the ear is least sensitive.[8]

CDs also use other forms of noise shaping to help compress part of the signal information and to help reconstruct the compressed signal.

Noise shaping works with a fixed noise type. It works in theory with white noise but in practice it must work with "colored" noise because again real noise does not have a flat frequency spectrum. Some more complex noise shapers do work in theory with colored or correlated noise. Still in all these cases the noise remains statistically the same and then the system acts on it. We turn now to the more difficult case where the noise changes and the noise model must change to catch up with it.

5.3. NOISE CANCELLERS LEARN NOISE PATTERNS TO ANNIHILATE THEM

Sound noise travels through the air as waves. Cancel or dampen the waves and you cancel or dampen the noise. One wave can cancel a second wave if they are sufficiently out of phase so that the first wave's peaks and troughs match the second wave's respective troughs and peaks. That is the governing principle of a noise canceller. And it is much easier said than done because of the changing and often random nature of the noise waves.

The oldest noise cancellers are fixed rather than adaptive. They try to dampen the intensity of the noise waves rather than directly cancel their peaks and troughs. Windows and walls are simple but practical examples. So are "sonic hedges" of shrubs or trees that grow between neighbors. More personal examples are wax or plastic earplugs or even one's hands cupped over one's ears. The latter example offers a modern noise context for Edvard Munch's famous 1893 painting *The Scream.*

Fixed noise cancellers have a simple mathematical basis: Most try to cancel or mask or suppress the *average* level of noise in a given environment. This principle underlies the design of office bullpens and car walls and the thickness of home windows. Civil engineers expressly or impliedly try to calculate and suppress this average noise level when they design multimaterial walls that separate busy city freeways from homes and lawns just on the other side of them.

Adaptive noise cancellers are more ambitious. They do not just try to estimate an average noise level and cancel it. They try to track the changing noise source that makes up the average noise level— and then cancel it. So an adaptive noise canceller acts as a double-duty signal processor that ceaselessly tries to follow the noise and then ceaselessly tries to reduce or eliminate it. Chapter 1 described an adaptive noise canceller that lets a physician listen to the heart-beat of an unborn fetus through the abdomen of the pregnant mother. Air and helicopter pilots use more sophisticated adaptive noise cancellers in their headsets to cancel a variety of engine vibrations and rotor noise and other cockpit noises.

Most adaptive noise cancellers rely on a badly named but quite useful algorithm called the LMS or "least mean square" algorithm. The LMS algorithm may be bit-for-bit the most important algorithm in modern signal processing and perhaps in all of signal processing.[9] Faxes and modems and satellite-based phone calls use LMS. So do hundreds of other information processing systems that need to track a noisy signal that changes in time. LMS also underlies modern adaptive antennas and so-called adaptive arrays and beam-formers.

The LMS algorithm is a simple neural network in which several input neurons feed into a single output neuron. The inputs flow over virtual wires or "weights" that resemble the tunable synapses that carry out learning among biological neurons. Bouts of feedback error correction modify the synapses in much the way that the behaviorist psychologist B. F. Skinner used operant conditioning and food pellets to modify the pecking behavior of pigeons. The LMS algorithm is simple because it is a linear system whereas modern neural networks are nonlinear. Still this compact and adaptive linear system remains the algorithm of choice for much real-world information processing because it is so easy to implement and so statistically powerful in its ability to track changing signals.

The Stanford electrical engineer Bernard Widrow developed the LMS algorithm in the late 1950s.[10] Norbert Wiener's earlier vision of "cybernetics" had started to morph into "artificial intelligence" or "AI." AI did and still does focus more on words and symbols than on numbers and statistics. Widrow kept his engineering focus on numbers and statistics and used LMS to create much of the modern field of adaptive signal processing as several awards attest. The Institute of Electrical and Electronics Engineers (IEEE) awarded him its prestigious Alexander Graham Bell Medal in 1986 for "fundamental contributions to adaptive filtering, adaptive noise and echo cancellation, and adaptive antennas." Widrow rejoined the neural networks research program in the 1980s and helped develop new generations of neural adaptive filters and noise cancellers. I helped organize the first IEEE international conference on neural networks that took place in San Diego in 1987—an event that many have seen as marking the start of the modern field of neural networks although the field was well under way by then.[11] We asked Bernie Widrow to give a plenary talk on the roots of the modern field of neural networks in the LMS algorithm. The tech media quickly drew a *Star Wars* analogy with the intellectual tension between the neural networks of the largely electrical-engineering community and the "AI" of the largely computer-scientist and robotic community—and dubbed Widrow the Obi-Wan Kenobi of neural networks. Then the IEEE gave Widrow its prestigious Third Millennium Medal in 2000. Widrow went on to win the 2001 Benjamin Franklin Medal in Electrical Engineering from the Franklin Institute for his "pioneering work in adaptive signal processing as exemplified by the LMS algorithm, adaptive filters, adaptive control, adaptive antennas, noise cancellers, artificial neural networks, and directional hearing aids." There are no Nobel Prizes in information science or "electrical engineering." If there were then Widrow's work on the LMS algorithm would surely have garnered one.

Widrow's creation or discovery of the LMS algorithm has an important historical footnote. His Ph.D. student Ted Hoff was his junior coauthor on the 1960 conference paper that first publicly stated the LMS algorithm. So it was technically a joint discovery in terms of the public documentary record. That explains why many researchers still refer to the LMS algorithm as Widrow-Hoff learning. Hoff finished his Ph.D. in electrical engineering in 1962 and later joined the then fledgling Intel Corporation in 1968. There Hoff promptly coinvented the first microprocessor in 1968. He was the lead author on the famous 1974 patent (U.S. patent number 3,821,715) behind the first single-chip microprocessor–the Intel 4004 4-bit chip that had appeared earlier in 1971. The National Inventors Hall of Fame inducted Hoff in 1996 for his pioneering work on the microprocessor.

The power and simplicity of LMS stem from an informed guess.

An LMS system is a learning system that maps noisy inputs to desired outputs. The system tries to learn at each moment the *average* error of a complex statistical process that changes with time. The average error is not only changing but unknown. Each average value itself requires a fantastic level of probabilistic knowledge about whether infinitely many different outcomes will occur. We never have such statistical knowledge. We have in practice just one sample per time instant–just one noisy footprint with which to estimate the changing foot that left it.

The idea behind LMS is that the footprint is good enough. Just use the footprint to estimate the foot even though the foot could have left infinitely many different footprints at that instant. This is anything but an Aristotelian balance of the possible with the actual. A given signal or noise value can in theory be any one of infinitely many values or numbers at a given instant. So the possibilities are infinite. It is a constrained infinity because some unknown probability bell curve describes how those possible values can occur. LMS says replace all those probability-weighted possibilities with the bird in hand: Replace all the statistical possibilities with the actual value that you observe or measure or detect.

The LMS approach can fail catastrophically in any given instance because it amounts to generalizing from a single sample. We run the same risk when we say that all the persons of a given race behave in such-and-such way because the one person from that race whom we have met or have gotten to know behaved in that way. But *on average* the LMS technique works well so long as the underlying world of signal and noise does not change too fast and so long as the noise itself is not too impulsive as we saw in chapter 4 that it can be. Statisticians describe this property of accurate average behavior as "unbiasedness"–the estimate equals the right value on average. This property of average behavior carries with it the usual proviso that a statistician can easily drown in a lake that is on average only six inches deep. Still LMS tends to work in practice because at each instant the unknown average error signal tends to shift only slightly and because then a fresh measurement or footprint arrives that also tends to resemble the unknown average value. LMS lets a system track the shifting average by subtly or grossly changing the values of its synaptic weights. There is a variant called LMAD (least mean absolute deviation) that can still track signals in highly impulsive noise such as the Cauchy and other types of stable noise that we saw in the previous chapter.[12]

LMS changes the synaptic weights of an output neuron based on "supervised learning" or feedback error correction. Many learning algorithms use some form of feedback error correction. The next chapter will discuss how the technique can find the optimal level of noise in many nonlinear systems. So a brief primer is in order.

Supervised learning starts with an error. A particular error is the gap between what you or your supervisor wants and what you get:

Error equals desired behavior minus actual behavior. The supervisor or teacher says what you want just as a parent tells a child how to behave at the table or in cleaning up the child's room. Supervised learning tries to make changes that reduce this error. Mathematical schemes such as LMS work with numbers. They turn a numerical error into a squared error to avoid negative numbers. They assume enough probability structure to compute or at least estimate the average squared error. Then they use the differential calculus to minimize that average squared error. The symbols may look complicated but the process itself is almost trivial if you know basic calculus.

Consider first the observed error at any given moment. Suppose you or the "supervisor" want the system to emit a 3 at one second if you put the value 6 into the system as input. But instead the system emits a 5. This gives an error of -2 because $3 - 5 = -2$. Then supervised learning feeds back the error signal of -2 to slightly adjust the weights or synapses that jointly produced the output 5. Suppose in the next second the system should emit a 2 when given the input 6. You feed the value 6 into the system but it now emits a 1. That gives a positive error of 1 because $2 - 1 = 1$.

But the observed error is again just one footprint that the random foot could have left. Consider again the observed error of -2 that occurs when you want a 3 but get a 5. A value of -2 or rather a clump of values very close to -2 has a probability of occurring. This tiny clump of values may lie to the left of a bell curve centered at 0 as did the bell curves in chapter 4. Or the probability curve may wiggle in some complex way. The precise value of -2 technically has zero probability of occurring because it is a precise numerical value and because the probability bell curve is continuous. So we formally must work with a tiny interval of values centered at -2 to avoid this zero-probability degeneracy. For our purposes we can think of -2 as occurring with some nonzero probability.

The important point is that there is a probability curve describing all possible error values and that this entire curve describes the average error that we could have expected to observe at that instant. But this involves infinitely many error values and their probabilities while all we have available is the single in-hand or observed error value of −2 (or its square 4). Hence it is mathematically radical in the extreme to view the unknown average error as simply the one observed error value −2 and likewise for the average squared error— because this amounts to using one choice out of a continuum of choices and ignoring its associated probability weight in the bargain. But it makes good pragmatic sense to do so: If the only error value we have at a given moment is the observed value of −2 then what other values could we use without making them up? This choice also provably leads to good performance in a wide range of mathematical cases in the sense that LMS will eventually find the optimal set of weights or synapses for the filtering task. And the thousands of successful applications speak for themselves.

An LMS noise canceller works by sampling the noise both by itself and by sampling it combined with a signal. It builds a mathematical model that tracks the ambient noise as it changes. Then it subtracts this estimated noise from the measured data that contains both signal and noise. The result tends to be just the signal if all goes well. Real systems often produce a faint or strong background hum from the canceling noise effect. Classical frequency filtering tends not to work in these cases because much of the noise spectrum overlaps the signal's spectrum.

Consider again the task of trying to adaptively filter a fetus's heartbeat. This requires adaptively canceling the maternal heartbeat and other maternal noise in fetal electrocardiography. Signal processors Victor Solo and Xuan Kong describe the LMS setup involved:

> Suppose we desire to measure the fetal heart rate
> during labor or delivery. The fetal heart rate can be

measured from a sensor placed in the abdominal region. The signal will be made noisy or distorted
chiefly by the mother's heartbeat but also by fetal motion. The idea behind noise canceling is to take a direct recording of the mother's heartbeat and, after
judicious filtering of this signal, subtract it off the fetal
heart rate signal to get a new relatively noise-free fetal
heart rate signal.[13]

Again the key idea is that canceling physical noise corresponds to
the mathematical operation of subtraction.

A few symbols help explain how LMS noise cancellation
works. Let the symbol S denote the signal of interest such as the
beating of a fetus's heart or the voice message that a helicopter pi-
lot hears in his headphones. Let the symbol N denote the noise in
the background. Let the symbol $\hat{N}$ denote the model that LMS
builds of the changing noise background N. The learning system
samples or observes the raw mixture of signal and noise. Assume
that the noise N adds to the signal S when it corrupts the signal.
Then the system uses a microphone or other sensor to sample the
raw mixture of signal plus noise or $S + N$. We saw in the last chapter that there are more complex cases where the noise multiplies
the signal but we focus here just on the simpler case of additive
noise. The system uses a second microphone or sensor to sample
only the raw noise source N. This lets the LMS system build an
adaptive model $\hat{N}$ of the noise source. If all goes well then the noise
model $\hat{N}$ will reasonably approximate the actual noise source N:
$\hat{N} \approx N$. Then adding $-\hat{N}$ or the negative of the noise model should
approximately cancel out the real noise source N in the actual observed noisy signal:

$$S + N - \hat{N} \approx S$$

The result is just the noise-filtered or noise-cancelled signal S or some close approximation to it.

This simple adaptive noise-canceling scheme has again thousands of real-world applications and appears in many devices. LMS lets pilots speak to one another and to remote persons even as they sit immersed in the roaring and varying noise of jet engines or rotor blades. Race car drivers use LMS-based systems in their headsets to cancel the noise from their engines and from their competitors' engines. It can filter out the random interference of a blinking human subject when measuring electrical activity with an EEG or electroencephalogram. LMS also adaptively filters out echoes on most phone lines.

Several of the students in my adaptive signal processing class have used LMS and its variants to adaptively cancel noise in nontechnical environments. One student used it to cancel the noise in his apartment when the children at the school next door came outside for recess. Many others have used it to help two people listen to each other talk against a background of cafeteria noise or street noise or the loud and often impulsive noise one hears when standing on a bridge over the 110 ("Harbor") freeway during rush hour. One student convinced some expert skydivers to let him make several tandem buddy-wrapped dives in the name of science. LMS cancelled much of the rushing wind during free fall as the adrenaline-pumped skydivers spoke to each other.

LMS is far from a perfect signal processing algorithm. Again the LMS noise canceller can fail if the noise is too impulsive. Several related algorithms such as the aforementioned LMAD algorithm tend to perform better but no model may effectively cancel some of the extremely impulsive stable noise discussed in chapter 4. The approach also fails if the noise changes too suddenly and loses its time correlation. The algorithm needs to sample the noise and make a few computations at each time instant. The noise speed or changes cannot be substantially faster than this processing speed for much the same reason that you cannot safely drive a car so fast that the curves and changes in the road ahead exceed the speed of your eye-hand coordination. Another vexing problem is statistical independence or lack of correlation between the signal and noise and between the signal and the estimated noise. A large literature addresses these and other problems that affect noise cancellation.

Some of these approaches use what my McMaster University colleague Simon Haykin and I call ISP or "intelligent signal processing" techniques.[14] These "smart" techniques include nonlinear neural networks and fuzzy-rule-based systems and several other "soft" approaches to signal processing. The ISP techniques tend to make fewer assumptions about the world and the signals and noise that infest it than do standard statistical signal processing techniques such as LMS. They also tend to require less preprocessing of raw measured signals but at the cost of defying easy mathematical description or control. So these ISP approaches tend to be brainlike in the unintended sense that they work well in many cases but they do not give the engineer an "audit trail" to follow that shows why they worked well. Neural networks are especially prone to this lack of transparency. They can quickly learn to track a complex and even impulsive noise process but they seldom come with a mathematical guarantee that they will do so within predefined error bounds. Novel inputs can lead to novel and disastrous results. Neural networks also share with brains the property that they forget some of their older learned patterns as they sample and learn new patterns— and we don't know which old patterns they have forgotten unless we test for them.

Both ISP and standard signal processing techniques have found their way into the world of wireless communications. Again LMS routinely cancels noiselike echoes in phone lines that would otherwise make it quite hard to speak in real time with someone in a remote location. Most of these applications involve efforts to filter or cancel noise from the wireless communication. We turn next to using noise as part of the very structure of most of modern wireless communications.

5.4. DELILAH'S SECRET: WIRELESS SIGNALS CAN HIDE IN NOISE

Wireless signals shoot through our bodies and brains wherever we go on the surface of the earth. They surround us and penetrate us in great numbers and combinations as if they were a seething and hidden world of electromagnetic phantoms. Many of these wireless signals are the long radio waves that course through our tissues at a specific carrier frequency. They carry all the radio stations one can find on the dial and other signals as well.

More and more wireless systems use *spread spectrum* to securely bury their signals in noise. Spread spectrum makes it hard for a snoop to intercept the signal or for an enemy aircraft or a next-door neighbor to jam a transmission. It also lets several users transmit and receive secure signals over the same swath of frequencies and at the same time.

Applications of spread spectrum range from the cell phones that all too many of us use in public places and now in airplanes to the twenty-four orbiting military satellites of the Global Positioning System that locate cell phone users and help guide cruise missiles to their targets. The Federal Communications Commission authorized the first commercial use of spread spectrum (in the megahertz range) in 1985. The military has used various forms of it since World War II. Winston Churchill and Franklin Delano Roosevelt spoke on secure lines in 1944 with the help of spread-spectrum radio links.

Spread spectrum spreads a signal across such a wide swath of frequencies or bandwidth that it looks like faint white noise. The signal does not transmit on just one carrier frequency or the equivalent of one lane on a highway. Instead it uses thousands or millions of lanes. Shannon's information theory implies that this increased bandwidth can increase the transmission's output signal-to-noise ratio. Thus spread spectrum gives another instance of the signal-noise duality because an eavesdropper or other unwanted interceptor sees the user's signal as just faint white noise.

There are many spread-spectrum schemes and hybrids. *Frequency hopping* is the oldest and one of the most popular schemes. It randomly jumps or "hops" the signal or a coded version of it from one frequency to the next. A synchronized receiver knows the hopping sequence and can reconstruct the hopped signal. The military uses frequency hopping to securely control remote sensing and weapons systems such as the unmanned aerial vehicles that fly over Afghanistan or Iraq.

Suppose we want to send a message in Morse code on a pipe organ. Morse code uses the binary symbols of a dot "." and a dash "–" to encode alphabetical letters and other message units. Morse code is a fairly efficient variable-length code because it assigns the shorter code words to the more frequently occurring symbols. It assigns the code word "." to the letter "e" while it assigns the code word "––.." to the letter "z" and so forth. Suppose we let a ticking stopwatch or metronome keep time with a steady uniform pulse rate. Then we can send Morse-coded messages using just the middle-C key on the organ if for each clock tick or time cycle we hold the key down briefly for a dot and hold it down longer for a dash. This scheme transmits digital information on the single carrier frequency that corresponds to middle C or about 261.63 hertz or cycles per second.

We can send the same message by a crude spread-spectrum method if we use different keys on the organ. We frequency-hop the message if we send the same sequence of dots and dashes (1s and 0s) but on randomly selected keys. The result would spread the signal over a fairly wide range of frequencies. It would also sound a little like noise or at least sound like "modern" atonal or pointillist music if the keystrokes went by fast enough and if the message was long enough.

Such a frequency-hopping scheme requires a sequence of random frequencies. Some real systems take samples from the thermal noise of resistors or other electronic devices in order to produce a nearly random or unpredictable sequence. There are also several computer algorithms that can generate such "pseudorandom" sequences. Most of these sequences still repeat after a given length but before that the sequence values tend to look as if they were samples from a white-noise process. I mentioned in the previous chapter that the U.S. Patent and Trademark Office issued a patent to me and a former Ph.D. student for an adaptive fuzzy system that comes up with if-then rules that in turn generate a white-noise-looking random sequence.[15] The rules slowly change in time. That makes it hard for an eavesdropper to detect the full sequence and makes it even harder for him to figure out what rules produced any given sequence values. There is here as elsewhere in modern communications a vast literature on alternative techniques.

Something simpler will suffice for the organ example. We could pick the next key to hit at random if we reached into a hat or bingo hopper and pulled out a ping-pong ball with a key name written on it. Then we would put the ball back into the hat or drum and rotate the drum and repeat until we had created a random-looking or pseudorandom sequence of keys. This takes time but we can do this in advance at our leisure. Then a distant receiver would have enough time to get the pseudorandom hopping sequence and listen to the dots and dashes on the hopped frequencies as they arrived if the sender and receiver synchronize their clocks. The receiver could "despread" the noisy-looking spread signal and recover the original Morse-coded message. The transmission itself would sound as if a frenzied cat jumped at random on the organ keys.

The frequency hopping would look even more like white noise if the organ had more keys. Imagine an electric organ with hundreds or thousands or even millions of keys. The dots and dashes could still randomly hop or jump from key to key across a much wider band of frequencies. The system would again require that a pseudorandom hop or key sequence use all the keys and that the sender and receiver synchronize their clocks. A computer would have to press the right keys at the right time. The human ear could not hear the high-frequency sound from most of the keys. The hopped signal would look still more like noise if we coded each message dot or dash and sent the coded message bits.

Spread spectrum can also arise through a *direct sequence* approach. Suppose we want to send the 3-bit message "1 1 0." The first step in direct-sequence spread spectrum converts the binary string into a bipolar string by replacing the 0s with −1s. This gives "1 1 −1." The next step replicates each bit several times. Suppose we replace each 1 with five 1s and replace each −1 with five −1s. This gives the redundant 15-bit bipolar string as follows:

redundant bipolar string = 1 1 1 1 1 1 1 1 1 1 −1 −1 −1 −1 −1

The next step takes a pseudorandom string of 1s and −1s and multiplies the paired-off 1s and −1s of the redundant bipolar string. A simple pseudorandom or noiselike string might be "1 −1 −1 1 −1" repeated three times to give the 15-bit string:

pseudorandom spreading string =
1 −1 −1 1 −1 1 −1 −1 1 −1 1 −1 −1 1 −1

The last step multiplies the two bipolar strings element by element to give a new noiselike spread string for transmission. The first bit of each string is 1. So $1 \times 1 = 1$ and thus the first bit of the new spread string is 1. The second bit of the redundant message string is 1 and the second bit of the pseudorandom string is −1. So $1 \times -1 = -1$ and thus the second bit of the new spread string is −1. Do this fifteen times. The result is the final noiselike spread string for transmission:

transmitted spread string = 1 −1 −1 1 −1 1 −1 −1 1 −1 −1 1 1 −1 1

Note that the first 10 bits of the final string are the same as the first 10 bits of the pseudorandom string. The last 5 bits are the bipolar opposites of the last 5 bits of the pseudorandom string. More random-looking spreading strings will produce more noiselike transmitted spread strings.

The sender now transmits the final string over a noisy "wireless" channel to a receiver. Real systems may send several such spread signals simultaneously. The receiver takes into account the time lag in the transmission to synchronize the despreading system. The despreading system reverses the process to recover the original message "1 1 0." This involves many other steps and banks of so-called matched filters that light up like light bulbs when either the appropriate signal or some close version of it arrives as a faint energy blip. It also involves sophisticated coding theory and cryptology. A fair amount of the research remains "black" or classified for security purposes.

Spread spectrum itself grew out of military projects in World War II and the cold war—and arguably from the innovation of the Hollywood film actress Hedy Lamarr.

The Lamarr story has become one of the great legends of twentieth-century electrical engineering. Hedy Lamarr was the strikingly beautiful brunette who played Delilah to Victor Mature's chubby Samson in Cecil B. DeMille's classic 1949 film *Samson and Delilah*. Figure 5.2 shows a publicity photo from the height of her celebrity. Lamarr had earlier appeared nude in the 1933 Czech film

Ecstasy. She was born Hedwig Eva Maria Kiesler in Vienna in the year 1913 or 1914. She took the stage name of Hedy Lamarr when Hollywood tycoon Louis B. Mayer signed her with MGM Studios in 1937.

Hedy Lamarr married first husband Friedrich Mandl when she was in her late teens. Mandl was an Austrian arms dealer. She was Jewish and so was Mandl. Yet Mandl did business with the Nazis and socialized with them. Young Hedy accompanied him to some of these gatherings and likely met Hitler and other Nazi bosses and may well have participated in their discussions of the latest weapons systems. Hedy ran away from Mandl in 1937 and went to London and then on to fame and glory in Hollywood. Mandl later lived in exile in Argentina. The IEEE History Center describes his involvement in the arms trade this way: "Mandl specialized in shells and

Figure 5.2: Film actress Hedy Lamarr (born Hedwig Eva Maria Kiesler in 1913 or 1914). Lamarr coauthored the 1942 U.S. patent "Secret Communication System" on frequency-hopping spread spectrum.

grenades, but from the mid-thirties on he also manufactured military aircraft. He was interested in control systems and conducted research in the field. Mandl kept his young wife by his side as he attended hundreds of dinners and meetings with arms developers, builders, and buyers, where Lamarr clearly learned some things."

So the question arises: How could Hedy Lamarr have come up with the sophisticated engineering idea of frequency-hopping spread spectrum? She had no formal training in engineering or science or mathematics. But she did have exposure to Viennese and Nazi arms experts and merchants. Did she hear or steal something from them? If so then the Nazis should have had access to the same technology and there is little or no evidence of that. If not then did she just have a sudden flash of technological insight?

The indisputable documentary evidence is that Hedy Lamarr coauthored the first U.S. patent on frequency-hopping spread spectrum. The Patent and Trademark Office issued patent number 2,292,387 to Hedy L. Markey and composer-writer George Antheil on 11 August 1942. Markey was the last name of Lamarr's second husband. The two coinventors had filed the patent the year before on June 10.

Hedy Lamarr's patent had the generic title "Secret Communication System." The patent proposed encoding the hop list of frequencies on what looks like a music roll from an old player piano. That presumably reflects the musical contribution from composer Antheil. The patent's first claim states that the frequency roll or "strip" will change the carrier frequency "from time to time in accordance with the recordings on said strip." It further proposes synchronizing the sender and receiver frequency scrolls. The patent suggests that the invention can securely guide a torpedo but leaves the door open to many other applications. Figure 5.3 shows one of the figures from Lamarr's patent and the proposed frequency scroll.

Reasonable doubt surrounds almost all else: Did Lamarr really

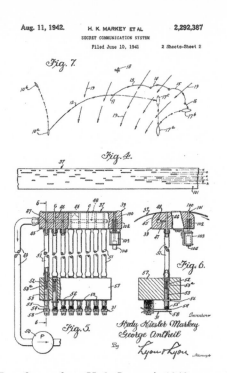

Figure 5.3: Four figures from Hedy Lamarr's 1942 patent for a "Secret Communication System" based on what we now call frequency-hopping spread spectrum. The etched scroll in figure 4 of the illustration encodes the random but synchronized hopping sequence for the frequencies.

come up with the idea of frequency hopping? Is it not more likely that she overheard something from her arms-merchant husband or his associates and that sparked the idea? How much of the idea did her composer coauthor contribute? How many unnamed sources helped Lamarr develop the idea because of her celebrity?

The evidence is equivocal. Other engineers did propose hopping-like schemes in the 1920s and 1930s but no one published anything comparable to what Lamarr and Antheil did. It also makes sense that the young Hedy Lamarr overheard something about anti-jamming possibilities when she sat with Mandl and his Nazi friends at dinner parties and the like. And again Lamarr lacked technical training even though she was both highly intelligent and creative. She never published anything else like her hopping scheme even though she and Antheil apparently toyed with some fanciful weapons designs before they had a falling-out.

Still the Patent and Trademark Office issued Lamarr and Antheil a frequency-hopping patent in 1942 based on the PTO's accepted and often quite rigorous procedures for claim scrutiny and searches of prior art. The PTO never revoked or modified Lamarr's patent as it could have done if it or challengers produced credible evidence of patent anticipation. So both the issuing of a valid patent and its term as a valid patent argue for a presumption in favor of Lamarr as prime innovator. The presumption is even stronger because the public record shows that many subsequent spread-spectrum patents have cited the Lamarr patent not just as prior art but as the founding patent for frequency hopping. The IEEE History Center makes a similar point: "Subsequent patents in frequency changing have referred to the Lamarr-Antheil patent as the basis of the field, and the concept lies behind the principal anti-jamming devices used today, for example, in the U.S. government's Milstar defense communications satellite system."

So on balance the documentary evidence of more than half a century supports the legend.

Hedy Lamarr died in 2000. The engineering world began telling her tale of innovation increasingly in the 1980s and 1990s as spread spectrum moved squarely into the commercial world of wireless communications. The popular media began to tell her tale much later in the 1990s as more and more news reporters and common folk logged on to the Internet and exchanged e-mails and sometimes e-traded stocks in the great stock market bubble at the end of the last millennium. I am personally aware of at least one effort to turn her patent story into a movie. Making such a film would not be a bad idea. The field of engineering could use some humanizing publicity given how for decades the media has stereotyped engineers as geeky social zeroes. It might also inspire more young people to study calculus and physics or simply to look more deeply into the digital world that surrounds us.

The Electronic Frontier Foundation gave the aged actress Hedy Lamarr its Pioneer Award in 1997 for her work on frequency hopping that she performed well over a half century earlier. Many newspapers around the world carried her cryptic response to the award: "It's about time."

CHAPTER 6

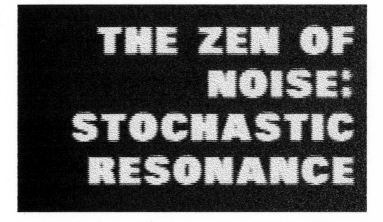

Stochastic resonance encapsulates the sexy notion that moderate (and carefully controlled) levels of noise in a nonlinear dynamical system can actually enhance the information throughput—and so improve the sensing and processing of otherwise undetectable signals. Originally postulated as a mechanism to explain how ice ages occur, the effect has since been demonstrated in a plethora of laboratory experiments and has also been proposed to be responsible for the way in which biological sensing mechanisms function to take advantage of inherent background noise.

—Adi R. Bulsara
"No-Nuisance Noise"
Nature, vol. 437, 13 October 2005

A functional magnetic-resonance imaging study applying a similar visual protocol indicates that neural activation in visual cortex appears improved by optimal noise.
   –Frank Moss, Lawrence M. Ward, and Walter G. Sannita
   "Stochastic Resonance and Sensory Information Processing"
   Clinical Neurophysiology, vol. 115, February 2004

Stochastic resonance aids detection of vibrating touch stimuli presented to the foot soles of both healthy elderly people with elevated vibrotactile thresholds and healthy young people with normal vibrotactile thresholds.
   –C. Wells, L. M. Ward, R. Chua, and J. T. Inglis
   "Touch Noise Increases Vibrotactile Sensitivity in
   Old and Young"
   Psychological Science, vol. 16, June 2005

The injection of noise induces chaotic dynamics triggering an El Niño [ocean warming] whenever the model's thermocline depth exceeds a threshold level.
   –Lewi Stone, Peter I. Saparin, Amit Huppert, and Colin Price
   "El Niño Chaos: The Role of Noise and Stochastic
   Resonance on the ENSO Cycle"
   Geophysical Research Letters, vol. 25, January 1998

A moderate amount of noise leads to enhanced order in excitable systems, manifesting itself in a nearly periodic spiking of single excitable systems, enhancement of synchronized oscillations in coupled systems, and noise-induced stability of spatial patterns in reaction-diffusion systems.
   –B. Lindner, J. Garcia-Ojalvo, A. Neiman, and
   L. Schimansky-Geier
   "Effects of Noise in Excitable Systems"
   Physical Reports, vol. 392, 2004

We might expect that the noise will "smear out" each data point and make it difficult for the network to fit individual data points precisely and hence will reduce over-fitting. In practice it has been demonstrated that training with noise can indeed lead to improvements in network generalization.

   —Christopher M. Bishop

   Neural Networks for Pattern Recognition

Increased noise in the transcription of a regulatory protein leads to increased cell-cell variability in the target gene output, resulting in prolonged bistable expression states. This result has implications for the role of noise in phenotypic variation and cellular differentiation.

   —James J. Collins et al.

   "Noise in Eukaryotic Gene Expression"

   Nature, vol. 422, 10 April 2003

Cells can use noise to suppress other noise or to create oscillations, multi-stabilities, and many other coherent kinetic traits.

   —Michael Springer and Johan Paulsson

   "Harmonies from Noise"

   Nature, vol. 439, 5 January 2006

Molecular motors operate by small increments, converting changes in protein conformation into directed motion.

   —Jeremy M. Berg, John L. Tymoczko, and Lubert Stryer

   Biochemistry

Rather than fighting it, Brownian motors take advantage of the ceaseless noise to move particles efficiently and reliably.

   —R. Dean Astumian and Peter Hänngi

   "Brownian Motors"

   Physics Today, vol. 55, November 2002

Noise can help as well as hurt. That theme has run throughout this book. We have seen that noise can cause hearing loss and sleep loss in humans and interfere with the mating songs of humpback whales. Acoustical noise often involves the common-law tort of private or public nuisance and underlies many zoning ordinances. Physical noise of many sorts and colors contaminates radar and sonar signals and our daily chatter over phone lines and wireless networks. But noise also can benefit some systems in the same sense that a small amount of poison can have beneficial effects while too much can harm or kill.

The first chapter gave a brief overview of the growing field of noise processing or stochastic resonance that seeks such noise benefits in the midst of nonlinear systems. This chapter looks deeper into why such noise benefits occur in neural systems and where they may lead. A concrete example is my own foray into stochastic resonance with a mathematical result called the "forbidden interval" theorem. This result guarantees noise benefits in many cases and led to predictions about how the retina likely uses noise to generate neural spikes. It also led to the first laboratory demonstration of a noise benefit in a carbon nanotube transistor. Such noise benefits can run much deeper in the molecular world of nanotechnology and quantum computing. Noise may even be the motive force of life.

6.1. MANY PHYSICAL AND BIOLOGICAL SYSTEMS DISPLAY A STOCHASTIC RESONANCE NOISE BENEFIT BECAUSE THEY ARE NONLINEAR SYSTEMS

Stochastic resonance means noise benefit. A system shows stochastic resonance or exhibits SR behavior if adding noise improves the system. The noise can be white or colored noise or chaos or any other energetic disturbance that interferes with some signal of interest present in the system. We can measure how well the system performs in terms of the system's signal-to-noise ratio or its input-output entropy or bit count or many other quantitative figures of merit.

The system itself must be nonlinear if it enjoys an SR noise benefit.

Linear systems do not benefit from noise because the output of a linear system is just a simple scaled version of the input: "The phenomenon does not occur in strictly linear systems where the addition of noise to either the system or the stimulus only degrades the measures of signal quality."[1] Put noise in a linear system and you get out noise. Sometimes you get out a lot more noise than you put in. This can produce explosive effects in feedback systems that take their own outputs as inputs. Then noisy inputs can destabilize the system much as when a wedding speaker puts his microphone too close to the speaker system. His microphone first picks up his amplified voice from the speaker system. Then the speaker system amplifies the voice again and feeds the amplified signal back to the microphone and round it goes until it ends in a loud screeching sound.

The lack of SR in linear systems also explains why so many engineers and scientists have overlooked the potential benefits of noise and indeed why they may find reports of SR effects surprising or counterintuitive. Many scientists and engineers work with linear models or with approximately linear models. At least much of their formal training likely focused on idealized linear systems where noise can only harm system performance. This tends to be truer of engineers than of physicists but it is still fairly common in the technical world. It is even more common in the world of basic textbooks because it is so comparatively easy to teach linear systems and to state and prove mathematical results about them. Linear systems can make great fodder for final exams. But research frontiers almost always involve nonlinear systems.

A threshold is the simplest nonlinear system. The threshold system turns on if the input value equals or exceeds some numerical threshold value and turns off it does not. Let the letter T denote a numerical threshold. We often take T to be zero but it can be any number. Let I denote the total input that arrives at the threshold system. A simple neuron model would view I as adding up all the on-off inputs that pour into the neuron at a given time from hundreds or tens of thousands of other neurons in the network. (The wires or synapses or connections between neurons can weight or multiply the signals that flow through them but we can ignore that complication here.) Then the system turns on or stays on if the input equals or exceeds the threshold (if $I \geq T$). The system turns off or stays off if the input falls short of the threshold (if $I < T$). The system is nonlinear because a small change in the input I can change the system output in the same way that just one more straw can break a camel's back.

Figure 6.1 shows how the pixels in a grayscale image can each act as lone threshold systems and produce an overall SR effect or noise benefit.[2] Engineers sometimes refer to this and similar noise effects as *dithering*.[3] The idea is that a faint signal can be too faint to cross a threshold and so adding noise energy can sometimes boost the signal above the threshold. The same structure arises in numerous physical and biological systems where a small amount of noise can nudge the system into a new state. Chapter 1 discussed some of the SR applications although there are hundreds of others and many show how noise can form and sustain complex patterns.[4] Chapter 3 also discussed how noise appears to benefit some auditory systems and cochlear implants. This chapter focuses on the special but important case of noise benefits in neurons.

Noise can benefit nervous systems because small amounts of it can improve how neurons process signals—or at least it does so in

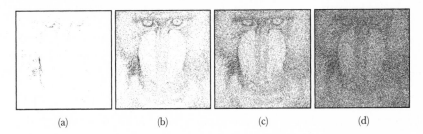

(a)   (b)   (c)   (d)

Figure 6.1: Stochastic resonance as pixel noise dither. The image in (a) shows the original and faint baboon image. Each image pixel acts as a simple on-off threshold neuron. Adding white Laplace noise to the pixels improves the image contrast and appearance in (b) and even more so in (c). Further noise degrades the image in (d).

almost all known models of neurons. Hundreds of researchers have published such SR noise benefits for model neurons. The models are sets of interlocking differential equations that describe how neurons change or evolve over time as they turn inputs into outputs. Other researchers have shown that adding external noise to the real neurons of crickets or rats or crayfish or other creatures improves the neurons' ability to detect signals or otherwise process information in noisy contexts. But these experiments still involve external noise that the scientist injects into the test subject.

The great open question of noise research is whether natural or *internal* noise sources in tissue will in fact produce an information benefit for real neurons or for networks of such neurons.[5] It would be no easy matter to locate internal sources of neural noise or other biophysical noise and then measure and track these noise effects on neurons. Again the theoretical evidence for such an SR effect is about as strong as it can be but that still is no substitute for an actual in vivo demonstration. The reward for such a demonstration may well be a Nobel Prize.

6.2. THE "FORBIDDEN INTERVAL" THEOREM: MODEL NEURONS BENEFIT FROM NOISE IF THE AVERAGE NOISE LIES OUTSIDE THE "FORBIDDEN INTERVAL"

We turn now to why SR noise benefits tend to occur in so many nonlinear models and especially in models of neurons or nerve cells. The simple answer is that these benefits occur because neurons behave at least approximately as thresholds and again thresholds can produce the noise benefit of stochastic resonance. The hundreds of billions of neurons in a brain either fire or not at any given moment of brain time. Each neuron emits a spike when it receives enough spikes from any of thousands of other neurons. Spike emission resembles a threshold process. The incoming spikes add up to trip the threshold much as enough water pouring into a dam can burst the dam. The actual spiking process involves a complex interplay between electricity and chemistry and millions of tiny ionic pumps. This is all evidence that thresholds can make reasonable models of neuronal spike emission. It does not explain why thresholds themselves produce an SR noise benefit.

Nor does it fully explain noise benefits in more complex models of neurons. Several biologists and physicists have observed some form of SR for mathematical models of randomly spiking neurons.[6] These models are more sophisticated than a simple threshold system because they endow a spiking neuron with some memory of its past actions. A simple threshold has no memory. It simply fires or not based on how many weighted signals flow into it at any given time. Still that simple memoryless model underlies the adaptive noise cancellers that we discussed in the previous chapter and helps reveal the structure of noise benefits in more general models of neurons and other nonlinear systems. That structure appears in a result we discuss below called the forbidden interval theorem.

Most SR results further require another assumption besides a threshold or thresholdlike nonlinearity between input and output. They require some form of a *subthreshold* signal.

Subthreshold signals model faint signals in nature or in digital systems. They are binary on-off signals that both lie below some threshold. The signals are binary in the sense that one signal value stands for yes or on or presence while the other signal value stands for no or off or absence. They are also usually bipolar as well as binary. This means that if a signal value A is a positive number such as 2 and stands for yes or on then the negative signal value $-A$ is -2 and stands for no or off. Such bipolar signals often occur in real communication systems when an energy blip stands for a 1 and its negative stands for a 0. The bipolar signals will be subthreshold if the threshold value T is larger than the signal value A. This gives the inequality relation $-A < A < T$. This inequality relation holds if A is 2 and if T is any larger number such as 4 because then $-2 < 2 < 4$.

Noise can help a threshold system detect signals in the presence of noise. Noise energy can boost the subthreshold signals closer to or above the threshold so that the threshold system can accurately decide whether the incoming noisy signal houses an on-signal (1) or an off-signal (0). This noise boost requires adding just the right amount of noise. Then the noisy version of the on-signal A tends to lie above the threshold T while the noisy version of the off-signal $-A$ tends to lie below the threshold T. So the threshold system will correctly turn on or "fire" or report a 1 on average when an on-signal arrives and will turn off or not fire or report a 0 when an off-signal arrives. Too much noise will push both the noisy on-signal A and the noisy off-signal $-A$ above threshold. Then the system cannot tell whether the incoming signal is an on-signal or an off-signal. Too little noise fails to trip the threshold at all and so the system again fails to distinguish on-signals from off-signals. Figure 6.3 below illustrates these three noise cases for realistic models of spiking neurons in the retina.

The forbidden interval theorem describes exactly when noise will help or hurt threshold systems with noisy subthreshold inputs. The theorem logically characterizes the SR noise benefit because it gives formal necessary and sufficient conditions for almost all noise types to benefit a threshold system or related thresholdlike neural system. That means noise will benefit the system if and only if the formal condition holds. The formal condition involves an interval of numbers. I have called this interval "forbidden" because the SR noise benefit occurs if and only if the average noise value does not lie inside this interval.[7]

Figure 6.2 shows the forbidden interval theorem at work. The first figure shows the signature "bump" or inverted-U curve of SR when we plot the noise intensity against the information or bit count of the threshold system. Noise at first increases the bit count and thus improves how the system performs when it tries to detect on-signals and off-signals in an incoming noisy stream of signals. The bit count rises to an SR maximum and then steadily falls as the noise increases beyond that point. The forbidden interval theorem predicts this result because the average noise level falls outside of the requisite interval that depends on the threshold value and on the subthreshold signal values. The right-hand graph in figure 6.2 shows no noise benefit because it shows that increasing the noise only decreases the system performance. The forbidden interval theorem predicts this outcome as well because in this case the average noise level lies inside the forbidden interval.

The forbidden interval theorem holds in great generality for almost all known types of noise if again two conditions hold: The system must be a memoryless threshold system and the signal inputs must be subthreshold in the sense that both the on-signal and off-signal lie below the system's threshold. Other noise benefits still occur without these assumptions. One example comes from SR researcher Nigel Stocks at the University of Warwick in England. Stocks has shown that networks of threshold neurons produce an SR noise benefit even when the inputs are not subthreshold and further that this benefit tends to increase with the number of neurons.[8] Stocks's results suggest that the aggregate behavior of a single neuron can still produce an overall SR noise benefit even when the signals are not all subthreshold. A very different example is the engineering work of French noise researcher François Chapeau-Blondeau at the University of Angers. He has shown that noise can still benefit many standard signal detection systems even though the optimal performance of such systems still occurs in the total absence of noise.[9]

The next question is whether the forbidden interval theorem still holds for more complex and more realistic models of neurons or other nonlinear structures. Most of these systems have some form of memory. That means the same input on one trial need not produce

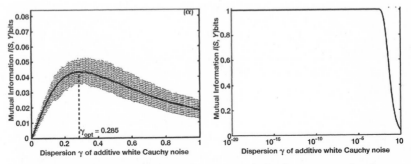

Figure 6.2: The forbidden interval theorem. A simple neuron or threshold system benefits from noise if and only if the average noise level falls outside the "forbidden" interval $(T-A, T+A)$ where T is the neuron's numerical threshold and where the subthreshold signals $-A$ and A obey $-A < A < T$. The left figure shows the signature inverted-U curve of a stochastic resonance noise benefit. This occurs because the location parameter of the impulsive Cauchy noise does not lie in $(T-A, T+A)$. The many vertical lines in the SR curve show the difference between the largest and smallest bit values when repeating the simulation several times. The right figure shows that noise only undermines the bit count of a threshold neuron when the average noise level falls inside $(T-A, T+A)$.

the same output as it did on an earlier trial unless enough of the previous inputs of the two trials are the same. Humans and other animals have memory in this sense. You need not get the same output or answer to the question "How do you feel today?" if you ask it of the same person in the same setting at different times. The internal system or dynamics change over time and thus so does the response to the same stimulus or to similar stimuli.

Weaker versions of the forbidden interval theorem still hold for many complex systems with memory. The theorems are weaker in the good sense that the forbidden interval condition is sufficient to produce an SR noise benefit but not necessary in general. SR noise benefits can still occur when the average noise falls inside the interval as well as outside of it. That is good news because it means SR noise benefits are more widespread in general than they are for simple thresholds without memory.

My Ph.D. student Ashok Patel and I showed that these more general noise theorems apply to biologically plausible models of spiking neurons.[10] These models view each neuron as random as well as having memory. A neuron can emit a spike at any given time but only with some probability. Several factors affect this spiking probability and differential equations govern the neuron's memory of its past input-output behavior. Noise has the potential to destabilize both the random spiking process and the feedback memory process. Too much noise will indeed overwhelm both processes. But small amounts of added noise can improve how the neuron fires on average or emits spikes given different inputs.

Figure 6.3 shows how background noise can help a spiking retinal neuron improve its firing rate in response to a brightness contrast image. The neuron responds to two random levels of brightness contrast in the top graph (a). One level is higher or brighter than the other. A coin flip decides if the brightness input at any given time is low or high. The neuron should emit the most spikes when the brightness

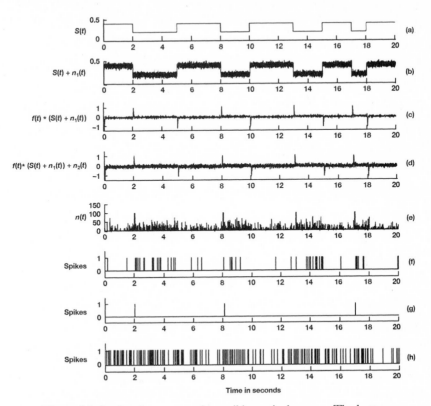

Figure 6.3: Stochastic resonance in a spiking retinal neuron. The bottom three rows show how a spiking retinal neuron responds to the random low-high contrast image in (a). The responses differ based on the level of additive white Gaussian noise as in (b). There are too few spikes in output (g) because the input noise level is too low. There are too many spikes in (h) because the input noise level is too high. Row (f) shows the SR effect where there is a nearly optimal number of spikes and in the right places because the input receives the optimal level of noise. The retinal spikes in (f) are most dense where the contrast pattern in (a) is lowest and least dense where it is highest.

level is low. It should emit far fewer spikes when the input brightness level is high. Again the process is random and so even an optimally firing retinal neuron will still fire a few times in the face of a high brightness level. The last three plots in figure 6.3 show that a small amount of noise improves the firing rate. Too little noise produces too few spikes while too much noise produces too many spikes.

The noise benefit of stochastic resonance raises a natural question: What is the *optimal* level of noise to add to a system?

This question has no easy mathematical answer because SR systems are sufficiently nonlinear to prevent direct analysis in most cases. The forbidden interval theorems also offer little insight here other than to act as a screening device. They can sometimes say whether a given system can benefit from any noise because they determine whether there is even a theoretical chance of finding a noise benefit in a given system. But the forbidden interval theorems do not say how to find such favorable noise. They ensure the existence of SR but they do not give a map or algorithm for finding the optimal level of noise if there is such a level.

That leaves the default approach of adaptation or neural learning. These techniques use training data and supervision to slowly change or adapt the system structure so as to move it toward a desired goal or state.

Adaptive systems learn by enlightened trial and error. The system can take a long time to learn well just as it can take a human a long time to learn to properly swing a golf club even with the help of the best golf instructor. But this iterative learning can also produce solutions that we could not find or at least could not find easily by pure mathematical analysis. I had worked with such learning systems for many years in the fields of neural networks and fuzzy systems. So it was only natural for my Ph.D. students and me to experiment with such algorithms. The resulting learning algorithms produce *adaptive* stochastic resonance.[11] Figure 6.4 shows some of the resulting learning curves. The noise intensities quickly converge to the optimal level of noise and then randomly jitter about that level.

SR learning algorithms may need millions of computer iterations to find the best amount of noise to add to the neuron or other nonlinear system. These adaptive algorithms add a small dose of random noise to the system and then let that noise churn through the system's often quite complex dynamics until the system emits an output value. The algorithm compares the output value with a desired value and computes an error signal much as an adaptive noise canceller does as we described in the previous chapter. The algorithm uses some form of that error value to slightly change one or more of the noise parameters and then it repeats the process. The algorithm will slightly increase the power or standard deviation or dispersion of the noise if the previous noise samples increased the system's bit count or increased its signal-to-noise ratio or any other specified performance measure. The algorithm is more complex than this because of the nonlinear nature of the system as the endnotes explain. The result eventually closes in on the optimal noise level as the three learning paths in figure 6.4 show. The same algorithms can also find the optimal level of chaos to add to a system to improve its performance.

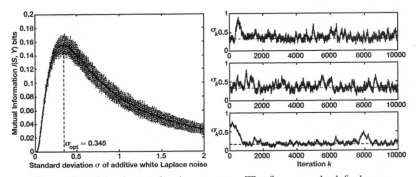

Figure 6.4: Adaptive stochastic resonance. The figure on the left shows the signature inverted-U signature of stochastic resonance (SR) for added white Laplace noise. The vertical line shows the optimal level of noise. The three learning curves on the right show how the adaptive SR system moves the noise strength or standard deviation until it locates the optimal level and then hovers about it in a type of random walk. The bottom learning curve shows a noise path where the noise strength starts out far too high and then quickly falls to the optimal level.

Adaptation suggests evolution or slow changes due to the variation and selection of system alternatives. The adaptive SR algorithms tend to find the optimal level of noise even when the noise is highly impulsive noise from a thick-tailed bell curve such as the Cauchy noise discussed in chapter 4. And again several scientists have documented some form of SR noise benefits in crayfish and paddlefish and crickets and other nerve-based creatures. Together these facts suggest that nature may have slowly stumbled on nerve and tissue structures that use noise to maximize the information in the local environment.

Nature may have taken millions or tens of millions of years to tune these wet parameters of signal processing just as it can take an adaptive SR algorithm millions of iterations before it locates the optimal noise level. But nature has had plenty of eons to find such biological settings in the gene pool. That is no guarantee that evolution did in fact tune our neural systems to exploit the many sources of ambient noise that affect neurons and neural computation. Still it is hard to believe that nature would not find or use neuron settings that would let neurons maximize their local information.

6.3. NOISE CAN BENEFIT NANOSYSTEMS AND THE MOLECULAR MOTORS OF LIFE

Nanotechnology is a new frontier for science and for noise. Nanosystems work at the level of a billionth of a meter. A nanoprocessor chip works with individual molecules while a microprocessor works with molecular structures a thousand times larger. But the nanorealm faces gusts of noise that can resemble gale winds from a hurricane. The random nature of the quantum world is itself a type of energetic noise when it interacts with some nanotech system of interest such as a nanochip or molecular gear or assembler. Noise at this level is built into the nature of things.

There is evidence that we can harness noise at the nanolevel in at least three broad areas. The first area is in nanotech devices such as carbon nanotubes. The other two areas are quantum computing and molecular motors.

Carbon nanotubes are cousins of diamonds because they are huge molecules of carbon atoms. Researchers in Japan isolated the first carbon nanotube in 1991 although humans have likely produced carbon nanotubes in cave soot for tens of thousands of years. A carbon nanotube resembles a roll of chicken wire or graphite with carbon atoms sitting snugly at the wire intersections. The cap at one end of the tube looks like half a soccer ball or half a so-called bucky ball or buckminsterfullerene of 60 tightly bound carbon atoms. The tube is about 100,000 times smaller in diameter than a human hair and can be fifty times as strong as steel wire of the same weight. Twisting the chicken-wire-like links between the atoms can produce a semiconductor or metal.[12]

Carbon nanotubes may be the best conductors of electricity that we have. They emit electrons and have started to appear in flat-panel displays. They also conduct heat well. We add nanotubes to some ceramics to make the ceramics harder. We also add nanotubes and related nanoparticles to golf balls to straighten their flight and to tennis rackets to give them more spring. Trillions of the tiny tubes can fit together in tiny arrays or carpets that one day may adorn household appliances or clothing both to strengthen the material and to smarten it by letting each tube or tube group act as a detector antenna or signal receptor. The trouble so far is that the tiny tubes tend to get tangled with one another. Future scientists and inventors will no doubt solve that and other problems in a variety of unforeseen ways. Noncarbon or inorganic nanotubes offer still other possibilities for nanoengineering as do long spun or woven nanotubes or nanowires and smaller quantum dots and related nanodevices.[13]

Nanotubes can also act as transistors or tiny logic gates in a nanochip. The journal *Science* named a nanotube transistor the break-through of the year 2001. Researchers soon showed that nanotube transistors could switch on and off in the microwave or gigahertz range of at least a billion times per second.[14] Several laboratories and companies continue to explore these possibilities and to file patents in the process.

I entered the nanotube gold rush with an obvious question for a noise researcher: What happens if we add a little electrical noise to a carbon nanotube transistor? I predicted that noise would boost the nanotube's performance.

The reasoning followed from the forbidden interval theorem and went like this: A nanotube transistor behaves like a threshold neuron when it emits an on- or off-signal. The simplest versions of the forbidden interval theorem state that any threshold system will show an SR noise benefit if the input signals are faint or subthreshold and if the average noise level does not fall in the forbidden interval of average noise values. The theorem is statistically robust in the sense that it still holds for highly impulsive or infinite variance noise as well as for the more common and well-behaved types of noise found in most textbooks. That robustness suggested that this SR theorem should also hold for the kind of soft or approximate threshold involved in a nanotube transistor. Thus some amount of noise should improve the performance of a nanotube transistor or on-off nanosignal detector because the nanotube transistor resem-bles a threshold neuron and because noise benefits all threshold neu-rons subject to the forbidden interval condition. So I predicted that a nanotube device would benefit from some level of noise if the device acted at least approximately as an on-off threshold. And thus a con-trolled experiment should either demonstrate or refute the predicted SR effect.

That proved far easier said than done. The noise prediction took only seconds to work out but the actual experiments took place off and on over a period of two years. I am a theorist and did not have any nanotubes or an atomic force microscope or any other equipment. Nor did I have the nanolab experience that the experiment would require. I did have a Ph.D. student named Ian Lee who had worked in industry with standard microprocessor transistors and who had recently worked with arrays of carbon nanotubes in a joint exploratory project with our colleagues at the Jet Propulsion Laboratory. So Ian was a good candidate to take the lead on this experiment and to make it the focus of his dissertation research. He was also willing to work the nights and weekends it might take to test and retest the noise prediction and then to start all over a few times and test and retest it again. All we needed were some nanotubes and a suitable laboratory.

I promptly proposed the noisy nanotube experiment to my USC colleague and nano-expert Chongwu Zhou. He was good enough to guide Ian and give him some nanotubes that he had grown and to let Ian have the off-hour run of his nanotech laboratory. This included access to an atomic force microscope that lets one manipulate nanotubes and other matter at the nanoscale. None of us would have agreed to do this if we had known how long it would take to control all the conditions and to isolate all the variables involved and then to repeat the experiments so that we would have enough data to make reliable statistical inferences—and to do it all on a shoestring budget that soon ran out.

Still the effort bore fruit. The data clearly showed the characteristic inverted-U of an SR benefit just as the forbidden interval theorems predict. That required extensive experimentation because again such theorems do not tell us where to look for the noise benefit or whether the noise benefit will be large enough for measurement devices to detect it. Adaptive SR techniques also did not help much because simulation models of the nanotube transistor either tended to omit so much structure that they were not realistic or they tended to include so many parameters to make them realistic that they were unreliable and nearly intractable. It took instead a great deal of trial and error to find a beneficial noise regime and to remove confounding variables under controlled conditions. But the noise benefit was there and it persisted even when we changed the measurement schemes and the types of interfering noise. The preliminary SR findings appeared late in 2003 in the journal *Nano Letters* of the American Chemical Society.[15] Future nanoengineers can use this information when they design nanotube signal detectors or other devices that operate in the presence of this type of noise.

Noise also holds promise to benefit the emerging field of quantum information processing. This new field has recast the "classical" information theory of Claude Shannon with its clever use of *qubits* or quantum bits in probabilistically entangled states. Quantum computers can in theory perform massively parallel computations that would exceed the capacity of standard computers and along the way crack a great many encryption schemes based on prime numbers. We already know that noise benefits quantum devices such as SQUIDs (supercomputing quantum interference devices) that detect faint magnetic fields.[16] Noise helps these systems change state along the lines of the usual SR metaphor in which a small amount of shaking can help a bouncing marble move from one well in an egg carton to another well. But it is a different question whether some form of noise will help quantum computers or quantum communication systems. A few papers have set forth theoretical reasons why we might expect an SR benefit in a quantum computer but so far we lack a formal theory or experimental demonstration.[17] That will likely change as the search for quantum SR benefits broadens.

The most intriguing nanolevel noise benefit may be to the origins of life itself.

Molecular motors drive motion in life processes that range from swimming bacteria and sperm to clenching our fingers into a fist.

These subcellular processes burn the basic biochemical fuel ATP (adenosine triphosphate) and use its enzyme form ATPase to catalyze the process and help steer it in just one direction. But the driving force of the molecular motor itself appears to be noise in the form of the thermal noise or Brownian motion that we discussed in chapter 4. The University of Edinburgh chemist David Leigh describes how molecular motors have made a virtue out of this noisy necessity: "For molecular sized motors inertia is negligible and the parts are subject to random and incessant Brownian motion. Rather than fight this effect biological motors use these random fluctuations in their mechanisms. Brownian motion drives both the power and exhaust strokes."[18] Other random motions may also help energize these molecular motors.

A motor turns energy into motion. A Brownian motor turns the energy of noisy thermal fluctuations into some form of useful work or movement. A showcase molecular motor is the mobile protein kinesin. It carries cargoes of proteins or other molecular particles in cells as it randomly "walks" in one direction along tiny microtubules or protein tracks. Kinesin can also help repair DNA sequences and perform several other cellular functions. Both kinesin and the RNA-synthesizing enzyme known as RNA polymerase appear to be Brownian motors. The Berkeley biochemist Carlos Bustamante explains that they "harness thermal fluctuations and rectify them using energy from chemical sources."[19] The chemistry component is crucial and means that there is still no overall free lunch.

Brownian motors or ratchets exploit random noise to move a gear or ratchet in one direction only. Chemical processes provide the one-way motion or rectification of the directionless Brownian motion. Random particles collide with a gearlike structure that can move in only one direction and so from time to time the gear advances by a notch. The net result is movement or propulsion in a preferred direction. The Caltech physicist Richard Feynman famously described an early version of such a thermal ratchet in his *Lectures on Physics*. Other physicists proposed similar ideas not long after Einstein's 1905 paper on Brownian motion. Today a subfield of physics explores these thermodynamic results in so-called fluctuation theorems and their many applications to biochemistry and nanotechnology—applications that exploit noise fluctuations and yet do not create perpetual motion machines or otherwise violate the second law of thermodynamics. The second law states that on average the total entropy or disorder of a system increases over time if the system is isolated.[20]

So is noise the secret of life?

That is at least a plausible hypothesis. Biochemical evolution appears to have adapted to the constant assault of thermal noise by using it to build motive structures out of simple proteins. Primitive components of the cell hitched a statistical free ride of sorts on the thermal fluctuations that came out of the inherent random froth of the quantum world. The hundred trillion or so cells in our bodies still depend on those energetic random fluctuations. So noise animates us as well as irritates us.

Noise may not be the secret of life. But there may have been no life without it.

CHAPTER 1: THE WAR ON NOISE

1. Experts in signal processing equate signals with information carriers: "The term *signal* is generally applied to something that conveys information. Signals generally convey information about the state or behavior of a physical system, and often, signals are synthesized for the purpose of communicating information between humans or between humans and machines." A. V. Oppenheim, R. W. Schafer, and J. R. Buck, *Discrete-Time Signal Processing*, 2nd ed., p. 8, Prentice Hall, 1999. "Signals represent information." A. Ambardar, *Analog and Digital Signal Processing*, p. 1, PWS Publishing, 1995. "A signal carries information, and the objective of signal processing is to extract useful information carried by the signal." S. K. Mitra, *Digital Signal Processing*, 2nd ed., p. 1, McGraw-Hill, 2001. "Anything that bears information can be considered a signal. For example, speech, music, interest rates, and the speed of an automobile are signals. This type of signal depends on one independent variable, namely, time and is called a one-dimensional signal. Pictures, x-ray images, and sonograms are also signals. They depend on two independent spatial variables,

and are called two-dimensional signals." C.-T. Chen, *Digital Signal Processing: Spectral Computation and Filter Design*, p. 1, Oxford University Press, 2001.

2. Two or more competitors are in a Nash equilibrium if each competitor does his selfish best given how his competitors act. The mathematician John Forbes Nash Jr. defined this concept of non-cooperative equilibrium and proved that it always exists when he was just twenty years old. J. F. Nash Jr., "Equilibrium Points in *n*-Person Games," *Proceedings of the National Academy of Sciences*, vol. 36, pp. 48–49, 1950. Nash shared the 1994 Nobel Prize in Economics for this two-page paper. The Nash equilibrium depends fundamentally on the competitive context: "Thus an equilibrium point is an *n*-tuple [of strategies in an *n*-product of probability simplexes] such that each player's mixed strategy maximizes his payoff if the strategies of the others are held fixed." J. F. Nash Jr., "Non-Cooperative Games," *Annals of Mathematics*, vol. 54, pp. 286–295, 1953. The 2001 film *A Beautiful Mind* dramatizes Nash's life and his discovery of the Nash equilibrium but misstates this fundamental concept of modern economics and game theory: B. Kosko, "How Many Blondes Mess Up a Nash Equilibrium?" *Los Angeles Times*, 13 February 2002.

3. An environment of political correctness can lead an unbiased advisor to "invest" in his reputation by lying from time to time to keep others from thinking that he is biased: "In fact she [the social-scientist advisor] is not racist but she has come to the conclusion that affirmative action is an ill-conceived policy to address racism. . . . If the social scientist is *sufficiently* concerned about being perceived to be racist, she will have an incentive to lie and recommend affirmative action. But this being the case, she would not be believed even if she sincerely believed in affir-

mative action and recommended it. Either way the social scientist's socially valuable information is lost." S. Morris, "Political Correctness," *Journal of Political Economy*, vol. 190, no. 21, pp. 231–265, 2001.

4. Imperial Spain controlled "*platina*" (platinum) mining in South America but radiocarbon dating shows that native South Americans had mined platinum and fashioned jewelry out of it over a thousand years before: "In 1748, according to the report of a visiting Spanish navy officer, 'Several of the mines have been abandoned on account of the platina, a substance of such resistance that when struck on an anvil of steel, it is not easy to separate. Nor is it calcinable, so that the metal enclosed with this obdurate body could only be extracted with infinite labor and charge.' " J. St. John, *Noble Metals*, p. 118, Time-Life Books, 1983.

5. The mother's heartbeat is the major source of noise: "The background noise due to muscle activity and fetal motion, however, often has an amplitude equal to or greater than that of the fetal heartbeat. A still more serious problem is the mother's heartbeat, which has an amplitude 2 to 10 times greater than that of the fetal heartbeat." B. Widrow and S. D. Stearns, *Adaptive Signal Processing*, p. 334, Prentice Hall, 1985. The mother's heartbeat is the (uncorrelated) random noise $n(t)$ that corrupts the fetal heartbeat signal s in the ECG measurement $s+n$. An adaptive noise canceller learns the time-varying signal s and approximates the noise n and then it subtracts the approximation y from the measurement to give the random error $e=s+n-y$. An ideal noise canceller minimizes the expected squared error $E[e^2]$ at each moment in time. But this expectation requires that at each instant the engineer have access to the underlying *unknown* probability density function that describes the likelihood of all possible signal-noise

combinations. The adaptive canceller in effect estimates this density afresh with each new measurement.

6. The paternity of information theory is not in dispute: "It is no exaggeration that Claude Shannon was the Father of the Information Age and his intellectual achievement one of the greatest of the 20th century." S. W. Golomb, "Claude E. Shannon (1916–2001)," *Science*, vol. 292, p. 455, 20 April 2001. "Information theory is one of the few scientific disciplines fortunate enough to have a precise date of birth. This special commemorative issue of the *IEEE Transactions on Information Theory* celebrates the 50th anniversary of Claude E. Shannon's 'A Mathematical Theory of Communication,' published in July and October 1948." S. Verdu, guest editor, "Information Theory: 1948–1998 Special Commemorative Issue," *IEEE Transactions on Information Theory*, vol. 44, no. 6, p. 2042, October 1998. "Information theory is one of those rare scientific fields to which one can assign a definite beginning. The publication in 1948 of Claude E. Shannon's paper 'A Mathematical Theory of Communication' marks the birth of information theory as clearly as the Declaration of Independence in 1776 marked the birth of a country." A. Ephremides and J. Massey, "1948–1998 Information Theory: The First Fifty Years," *IEEE Information Theory Society Newsletter*, p. 1, Summer 1998.

7. Gauss first discovered many important mathematical properties that scientists have only recently rediscovered. The FFT is just one example: M. T. Heideman, D. H. Johnson, and C. S. Burrus, "Gauss and the History of the Fast Fourier Transform," *IEEE ASSP Magazine*, vol. 1, no. 4, pp. 14–21, October 1984.

8. The physicist Paul Davies uses Stephen Hawking's black hole information structure (which in turn uses Shannon's entropy

measure) to derive the bit count of the universe: "One may obtain a natural measure of I the information capacity of the cosmos using the Hawking-Bekenstein formula for black hole entropy. If the entire universe were converted into a black hole, it would conceal a quantity of information I given by $I \approx \dfrac{Gm^2}{\hbar c}$ where m is the mass of the observable universe (i.e., within the particle horizon). At the current epoch t_0, $I \approx 10^{120}$. At epoch t: $I(t) \approx 10^{120} \left(\dfrac{t}{t_0} \right)^2$ so that at the Planck time $t_p \approx 10^{-43}$ s, $I \approx 1$ as expected." P. C. W. Davies, "Why Is the Physical World So Comprehensible?" in *Complexity, Entropy, and the Physics of Information: Proceedings of the Santa Fe Institute in the Sciences of Complexity*, ed. W. H. Zurek, vol. 8, p. 5, Addison-Wesley, 1990.

9. A half century of neural research finds that neural spikes are rich sources of information singly as well as jointly in spike trains: "Individual spikes are important. In the billions of neurons that are active as you read this text, each firing perhaps tens of spikes per second, it is difficult to believe that one spike more or less could matter. Yet we have seen [in the chapters of this book] that, under many conditions, behavioral decisions are made with of order one spike per cell, that individual spikes can convey several bits of information about incoming sensory stimuli, and that precise discrimination could, at least in principle, be based on the occurrence of individual spikes or spike pairs at definite times. These different results encourage us to take seriously the possibility that each spike that streams into our brain really does make a difference." F. Rieke, D. Warland, R. de Ruyter van Steveninck, and W. Bialek, *Spikes: Exploring the Neural Code*, p. 279, MIT Press, 1997.

10. Bits through queues can "leak" information through timing noise: "Timing can indeed be used to subvert security boundaries in a communication network by leaking information through the

timing of packets while ostensibly sending only innocuous information. It can also be used to provide hidden channels for collu-
sion among agents involved in a negotiation over a network." V.
Anantharam and S. Verdu, "Reflections on the 1998 Information
Theory Society Paper Award: Bits through Queues," *IEEE Information Theory Society Newsletter*, vol. 49, no. 4, pp. 20–24, December
1999. The authors derive the channel capacity of an exponential
queue in their pioneering paper "Bits through Queues," *IEEE
Transactions on Information Theory*, vol. 42, no. 1, January 1996.
They show that the capacity of an exponential single-server queue
is the smallest of any service probability distribution and thus the
exponential queue is the noisiest of all such queues. Its capacity
is e^{-1} nats or 0.531 bits per average service time. A like analysis
finds the capacity in a discrete geometric queue: A. S. Bedekar
and M. Azizoglu, "The Information-Theoretic Capacity of
Discrete-Time Queues," *IEEE Transactions on Information Theory*, vol.
44, no. 2, March 1998. For a historical perspective of the bits-
through-queues breakthrough see A. Ephremides and B. Hajek,
"Information Theory and Communication Networks: An Uncon-
summated Union," *IEEE Transactions on Information Theory*, vol. 44,
no. 6, October 1998.

11. T. M. Cover and J. A. Thomas, *Elements of Information Theory*, p.
183, John Wiley & Sons, 1991.

12. We define the mutual information $I(X, Y)$ as the (pseudometric)
Kullback "distance" between the joint probability density $p(x, y)$
and the product density $p(x) p(y)$: $I(X, Y) = \sum_x \sum_y p(x, y) \log_2 \frac{p(x, y)}{p(x)p(y)}$
where infinite-ranged but finite integrals replace the sums for
continuous probability densities. Then mutual information
equals receiver entropy minus the receiver's conditional entropy:
$I(X, Y) = H(Y) - H(Y|X)$ where the unconditional entropy is

$H(Y) = -\sum_y p(y)\log_2 p(y)$ and the conditional entropy averages over the source density $p(x)$ to give $H(Y|X) = \sum_x p(x)H(Y|X=x)$ where $H(Y|X=x) = -\sum_y p(y|x)\ \log_2\ p(y|x)$. Jensen's inequality and the concavity of logarithms imply that $I(X, Y) \geq 0$. So $H(Y) \geq H(Y|X)$ and thus the information-theory slogan that "conditioning reduces entropy." The easy and important fact that the text tries to explain in words alone is that the mutual information also equals the receiver entropy minus the receiver's conditional entropy with a symmetric result for the source: $I(X, Y) = H(Y) - H(Y|X)$. The channel capacity maximizes this entropy gap or information gain over all possible source probability densities $p(x)$: $C = \max\ I(X, Y)$. The capacity of the binary symmetric channel in figure 1.1 is $C = 1 - H(p)$ where $H(p) = -p\ \log_2\ p - (1-p)\log_2 - (1-p)$. Hence this capacity is zero bits if $p = \frac{1}{2}$ and thus if any transmitted bit is as likely as not to flip from 1 to 0 or from 0 to 1. This capacity is 1 bit if such noisy-channel bit flipping is certain $(p=1)$ or if it is impossible $(p=0)$.

13. Shannon derived a logarithmic relationship to describe the channel capacity of a bandlimited power-constrained Gaussian channel that has a signal power of P watts and a noise spectral density of $N/2$ watts per hertz: $C = W\ \log_2(1 + \frac{P}{WN})$ bits per second. Hence this capacity converges to $C = \frac{P}{N}\ \log_2 e$ bits per second as the bandwidth W grows to infinity (also giving the 1-bit value of .693 kT joule). Hence infinite bandwidth gives only a finite channel capacity but one that depends in a linear way on the signal power P. For details see chapter 10 of T. M. Cover and T. A. Joy, *Elements of Information Theory*, John Wiley & Sons, 1991. But fiber optics integrates the nonlinear stochastic Schrödinger wave equation to describe an optical fiber's input-output characteristic. The integral produces *multiplicative* noise that destroys the linear growth in the above channel capacity: "The reason for the non-monotonic

behavior of the capacity is that if we consider a particular channel, the signal in the other channels appears as noise in the channel of interest, owing to the nonlinearities. This 'noise' power increases with the 'signal' strength, thus causing degradation of the capacity at large 'signal' strength." P. P. Mitra and J. B. Stark, "Nonlinear Limits to the Information Capacity of Optical Fiber Communications," *Nature*, vol. 411, pp. 1027–1030, 28 June 2001.

14. The universe will expand forever and not contract if it is "flat" or Euclidean. That is just what a team of several astrophysicists found and reported as "the first images of resolved structure in the microwave background anisotropies over a significant part of the sky. . . . This is consistent with that expected for cold dark matter models in a flat (Euclidean) universe." P. de Bernardis et al., "A Flat Universe from High-Resolution Maps of the Cosmic Microwave Background Radiation," *Nature*, vol. 404, pp. 955–959, 27 April 2000. Measurements from NASA's more recent Wilkinson Microwave Anisotropy Probe (WMAP) satellite confirm the flatness and likely endless expansion.

15. J. P. Ostriker and P. J. Steinhardt, "The Quintessential Universe," *Scientific American*, pp. 47–53, January 2001.

16. T. Shinbrot and F. J. Muzzio, "Noise to Order," *Nature*, vol. 410, p. 251, 8 March 2001.

17. S. Mitaim and B. Kosko, "Adaptive Stochastic Resonance," *Proceedings of the IEEE*, vol. 86, no. 11, pp. 2152–2183, November 1998.

18. The following papers are good introductions to stochastic resonance: K. Wiesenfeld and F. Moss, "Stochastic Resonance and the Benefits of Noise: From Ice Ages to Crayfish and SQUIDs,"

Nature, vol. 373, pp. 33–36, 5 January 1995; J. J. Collins, "Stochastic Resonance without Tuning," *Nature*, vol. 376, pp. 236–238, July 1995; F. Moss and K. Wiesenfeld, "The Benefits of Background Noise," *Scientific American*, vol. 273, no. 2, pp. 66–69, August 1995; A. R. Bulsara and L. Gammaitoni, "Tuning into Noise," *Physics Today*, pp. 39–45, March 1996; L. Gammaitoni et al., "Stochastic Resonance," *Reviews of Modern Physics*, vol. 70, no. 1, pp. 223–287, January 1998.

19. One stochastic resonance theory sees noisy fluctuations in the level of freshwater in the ocean as the potential cause of ice-age periodicity: "We propose that the glacial ocean circulation, unlike today's, was an *excitable system* with a stable and a weakly unstable mode of operation, and that a combination of weak periodic forcing and plausible-amplitude stochastic fluctuations of the freshwater flux into the North Atlantic can produce glacial warm events similar in time, evolution, amplitude, spatial pattern, and interspike intervals to those found in the observed climate records." A. Ganopolski and S. Rahmstorf, "Abrupt Glacial Changes Due to Stochastic Resonance," *Physical Review Letters*, vol. 88, no. 3, pp. 038501-1–038501-4, 21 January 2002.

20. Both mathematical theorems and detailed simulations show that noisy spiking neurons can use many types of noise to adaptively maximize the Shannon mutual information between input and output spike trains. The general theorems state that all threshold systems (including neurons) exhibit stochastic resonance for all possible noise probability density functions with finite variance and for all stable infinite-variance noise: B. Kosko and S. Mitaim, "Stochastic Resonance in Noisy Threshold Neurons," *Neural Networks*, vol. 16, no. 5–6, pp. 755–761, June 2003. A converse occurs in B. Kosko and S. Mitaim, "Robust Stochastic Resonance for

Simple Threshold Neurons," *Physical Review E*, vol. 70, p. 031911 (10 pages), 27 September 2004. A special case occurs when a carbon nanotube acts as a thresholdlike antenna: I. Lee, X. Liu, B. Kosko, and C. Zhou, "Nanosignal Processing: Stochastic Resonance in Carbon Nanotubes That Detect Subthreshold Signals," *Nano Letters*, vol. 3, no. 12, pp. 1683–1686, December 2003. For evidence that general adaptive stochastic resonance is robust against even highly impulsive noise see B. Kosko and S. Mitaim, "Robust Stochastic Resonance: Signal Detection and Adaptation in Impulsive Noise," *Physical Review E*, pp. 051110-1–051110-11, 22 October 2001. These stochastic resonance effects persist in more biologically plausible neurons that possess memory dynamics: A. Patel and B. Kosko, "Stochastic Resonance in Noisy Spiking Retinal and Sensory Neuron Models," *Neural Networks*, vol. 18, pp. 467–478, July 2005.

CHAPTER 2: NOISE IS A NUISANCE

1. The Prosser quote comes from the famous torts "hornbook" or monograph from first-year law: W. Page Keeton, *Prosser and Keeton on Torts*, 5th ed., p. 616, West Publishing, 1984.

2. The court found that an improperly listed phone number in the yellow pages created so many unwanted phone calls as to count as private nuisance and then granted relief for the resulting emotional distress. The phone company had wrongly listed the plaintiff's personal phone number as the number of an "after hours" florist: "We conclude that the erroneous listing of plaintiff's telephone number and the numerous telephone calls to plaintiff resulted in an invasion of plaintiff's right to enjoy her property without unreasonable interference. As such it is governed by the

law relating to a private nuisance, and plaintiff is entitled to recover for mental distress resulting from defendant's negligent act."
Macca v. General Tel. Co. of Northwest, 262 Or. 414, 418; 495 P.2d
1193 (19 April 1972).

3. Fuzzy logic permits a formal fuzzification of the Coase theorem so
    that if the theorem's if-parts hold only partially then the conclu-
    sion holds partially but according to an exact mathematical for-
    mula. See chapter 6 of B. Kosko, *Heaven in a Chip*, Random
    House, 2000. Chapter 6 endnote number 18 states this metatheo-
    rem: "A fuzzy Coase Theorem with binary if-then strength has
    the *modus ponens* form

    $$If \quad t_L(A \to B) = 1$$
    $$and \quad t(A) \geq a$$
    $$then \quad t(B) \geq a$$

    because $c = 1$ [in $t_L(A \to B) = c$ where the Lukasiewicz multivalued
    implication operator has the form $t_L(A \to B) = \min(1, 1 - t(A) + t(B))$
    for truth values $t(A) \in [0,1]$ implies that $t(B) \geq \max(0, c + a - 1) = $
    $\max(0, a) = a$. Here the statement A stands for the if-part conjunc-
    tion 'Property rights are well-defined (binary) and transactions
    costs are zero.' The min operator can factor this statement into a
    truth function of its two conjuncts: $t(A) = \min(t(P), t(T))$ where P
    stands for the statement 'Property rights are well-defined (binary)'
    and T stands for 'Transactions costs are zero.' Product could also
    define the conjunction operator." Endnote 14 derives the more
    general case where the truth value c of the if-then conditional need
    not be unity.

4. W. Page Keeton, *Prosser and Keeton on Torts*, 5th ed., p. 622, West
    Publishing, 1984.

5. Low transactions costs let parties bargain away their disputes and thus invoke the Coase theorem: "When transactions costs are low, it is efficient for the parties to resolve incompatibilities on their own via consensual exchange . . . when transactions costs are high, we have seen that involuntary exchange may be more conducive to efficiency. In this case, courts should take a more active role by coercing exchange at a prescribed price." T. J. Miceli, *Economics of the Law: Torts, Contracts, Property, Litigation*, p. 118, New York: Oxford University Press, 1997. The text quote is from the same page.

6. Binary property boundaries can promote aggressive play while fuzzier boundaries can sometimes promote tolerance: "Some neighbors at least will yield to the temptation to act strategically if boundary lines are given absolute respect. . . . Even in the best of all worlds, it would cost money, and impose impediments and encumbrances on legal title, to negotiate thousands of transactions to reach the position where 'live and let live' places us from the start. In this setting, relaxing the sharp boundary makes perfectly good sense, and the relaxation should come in the only way possible, as a matter of law." R. A. Epstein, *Principles of a Free Society: Reconciling Individual Liberty with the Common Good*, p. 194, Reading, Mass.: Perseus Books, 1998. The text quote is from page 191.

7. W. Page Keeton, *Prosser and Keeton on Torts*, 5th ed., p. 627, West Publishing, 1984.

8. A. J. Casner, W. B. Leach, S. F. French, G. Korngold, and L. VanderVelde, *Cases and Text on Property*, 4th ed., pp. 893–897, Aspen Law & Business, 2000.

9. J. M. Glionna, "Strife After Death," *Los Angeles Times*, p. A1, 13 November 2003.

10. Horror movie fans and critics of cryonics often grossly overesti-
mate the actual number of persons who have signed up for cryonic suspension and the still smaller number of those actually
suspended in liquid nitrogen: "The $25-billion funeral industry
opposes cryonics even more vigorously than it opposed cremation
before the 1960s. It buries about 6,000 people every day in expensive coffins while only a thousand or so folks have signed up
for cryonics and only about 100 actually lie in cryo-suspension."
B. Kosko, "Despite Skeptics and Critics, Cryonics May be a Cool
Way to Go," op-ed, *Los Angeles Times*, 19 July 2002.

11. The excerpt comes from the dissenting opinion of Pennsylvania
Supreme Court chief justice (and former governor of Pennsylvania
for just 19 days) John Cromwell Bell Jr. in *Griggs v. Allegheny
County*, 402 Pa. 411, 422 (1961); 168 A. 2d 123, 128–129 (Pa. 1961).

12. Louisville has moved hundreds of families away from its noisy
airport: "Moving an entire city and most of its 552 families is a
first for an airport. The FAA has recognized the approach as in-
novative and is financing the $30-million first phase of the Her-
itage Creek project with matching funds from the airport industry."
J. Ott, "Airport Moves City Away From Noise," *Aviation Week &
Space Technology*, p. 66, 9 August 1999.

13. Legal commentators have pointed out that the new doctrine of virtual trespass to chattels seems to borrow as much from the old tort
of trespass to land as from the tort of trespass to chattels: "Courts
have been able to overlook the trespass to chattels' harm requirement rather effortlessly because there is no such requirement for real
property. If an actor trespasses on another's land, the owner does
not have to prove any harm to recover. The trespasser is strictly li-
able. By allowing the operators of computer and Internet systems to

obtain injunctions under the trespass-to-chattels doctrine without demonstrating that the systems have suffered physical harm, courts have acted as if the operators are owners of real property." J. Beauregard, "*Intel Corp. v. Hamidi*: Trespassing in Cyberspace," *Jurimetrics*, vol. 43, p. 490, 2003. The text quote comes from the same page.

14. Law-and-economics scholar Richard Epstein argues that it makes more sense to view cyberspace as virtual land than as virtual chattels or personal property: "Common language speaks of Internet 'addresses,' for, of course, individuals and firms occupy private 'sites' along the Internet 'highway.' It also speaks of the 'architecture' of the Internet . . . [A]n Internet site is fixed in its cyberspace location. To change from one address to another risks the loss of its customer base, just like any ordinary store runs the risk of losing its customers when it changes locations. In these circumstances, cyberspace looks and functions more like real property than chattels." R. A. Epstein, "Cybertrespass," *University of Chicago Law Review*, vol. 70, pp. 82–83, 2003.

Other legal scholars have challenged the cyber-analogy with either personal or real property: "Unlike land and chattels, a Web site is intangible property containing only information. The public interest in access to information on Web sites is likely to be greater than its interest in accessing another's land or chattels. There are First Amendment free-speech concerns that should be weighed before the law grants rights to control access to Web sites. Trespass does not account for such interests." M. A. O'Rourke, "Is Virtual Trespass an Apt Analogy?" *Communications of the ACM*, vol. 44, no. 2, pp. 101–102, February 2001. The California Supreme Court accepted much of this skepticism in its 2003 decision in *Intel v. Hamidi*: "The California Supreme Court . . . decided that meaningful harm to a computer system must be shown before such a trespass claim can succeed. The California court recognized that stretching trespass law to stop unwanted but harmless email could interfere with free speech and other socially valuable interests." P. Samuelson, "Unsolicited Communications as Trespass?" *Communications of the ACM*, vol. 46, no. 10, p. 15, October 2003.

15. D. L. Burk, "The Trouble with Trespass," *Journal of Small and Emerging Business Law*, vol. 4, no. 1, p. 30, Spring 2000.

16. Legal scholar Adam Mossoff argues that "spammers are creating negative externalities through the use of their e-mail accounts. Nuisance doctrine is an ideal legal mechanism for forcing spammers to internalize the costs they impose on innocent third parties." A. Mossoff, "Spam—Oy, It's Such a Nuisance!" *Berkeley Technology Law Journal*, vol. 19, p. 665, Spring 2004. The text quote is from page 650.

CHAPTER 3: THE NUISANCE THAT DEAFENS

1. A. S. Niskar, S. M. Kieszak, A. Holmes, E. Esteban, C. Rubin, and D. C. Brody, "Prevalence of Hearing Loss Among Children 6 to 19 Years of Age," *Journal of the American Medical Association*, vol. 278, pp. 1071–1075, 1998.

2. P. U. Teie, "Noise-induced Hearing Loss and Symphony Orchestra Musicians: Risk Factors, Effects, and Management," *Maryland Medical Journal*, vol. 47, no. 1, pp. 13–18, 1998.

3. These noise facts come from the National Institute for Occupational Safety and Health's Web site and from its summary document "Work-Related Hearing Loss" at http://www.cdc.gov/niosh/hpworkrel.html.

4. A decibel measures sound with scaled base-10 logarithms. The scale factor will be 20 or 10 depending on whether the decibel measure takes the logarithm of a ratio of squared quantities such as *sound pressure* or whether it takes the logarithm of nonsquared quantities such as either the *sound power* in watts or the *sound intensity* given as sound power in watts per square meter. The decibel examples in the text use the former measure for sound pressure: "Psychophysical experiments have established that we perceive an approximately equal increment in loudness for each 10-fold increase in the amplitude of a sound stimulus. . . . The level L of any sound may then be expressed (in units of decibels sound-pressure level or dB SPL) as

$$L = 20 \log_{10}\left(\frac{P}{P_{ref}}\right)$$

where P, the magnitude of the stimulus, is given as the root mean square of the sound pressure (in units of pascals or PA). For a sinusoidal stimulus, the peak amplitude exceeds the root mean square by a factor of the square root of 2. As the reference level on this scale, $0\,dB$ SPL is defined as the sound pressure whose root mean square value is $20\,\mu Pa$. This intensity corresponds to the approximate threshold of human hearing at $4\,kHz$, the frequency at which our ears are most sensitive." E. R. Kandel, J. H. Schwartz, and T. M. Jessell, *Principles of Neuroscience*, 4th edition, p. 593, McGraw-Hill, 2000. So $60\,dB = 20 \log_{10} R$ for ratio of sound pressures R. Division and exponentiation gives $R = 10^3$ or a factor of a thousand. Similarly $120\,dB$ gives $R = 10^6$ or a factor of a million while $180\,dB$ gives $R = 10^9$ or a factor of a billion.

5. Helmholtz's 1877 magnum opus assumes that music is periodic sound (its appendix includes a derivation of sine-wave solutions

to a simple differential equation): "Those regular motions which produce musical tones have been exactly investigated by physicists. They are *oscillations, vibrations,* or swings, that is up and down, or to and fro motions of sonorous bodies, and it is necessary that these oscillations should be regularly periodic." H. L. F. Helmholtz, *On the Sensations of Tone as a Physiological Basis for the Theory of Music,* p. 8, Dover Publications, 1954. The text quote comes from the same page.

6. The Russian portion of the Space Station has been the noisiest: "Readings taken shortly after Russia's service module was attached in July show that noise levels average more than 70 decibels. That would make it as noisy as a machine room or a rattling air conditioner." C. Holden, "Noisy Days Aboard the Space Station," *Science,* vol. 290, p. 2249, 22 December 2000.

7. A CD player attaches to the noise gun: "[Inventor Ellwood] Norris uses 50 different sound tracks, or sonic bullets, in his new weapon. . . . The noise level [140 decibels] is similar to that of a passenger jet taking off." P. Pae, "Weapon Sends Message That's Loud and Clear," *Los Angeles Times,* 23 June 2002.

8. The European Union's noise maps are three-dimensional and color-coded: "The maps are stunning, with average noise levels superimposed on 3D visualizations of entire cities in colored contours ranging from pale-green—less than 45 decibels—to deep blue for greater than 79 decibels. . . . When anecdotal evidence and parochial complaints once reigned, the public can now both better argue its own particular gripes and be convinced at a glance of the broader benefits of unpopular controls, such as stricter traffic speed limits." "Sound Thinking," editorial, *Nature,* vol. 427, p. 471, 5 February 2004.

9. Each inner-ear hair cell contains on the order of 100 ion channels that transduce mechanical energy into electrical impulses: "Signal-channel recordings and noise analysis suggest that each hair cell possesses only about 100 transduction channels . . . there is a comparable number of stereocilia in the hair bundle." E. R. Kandel, J. H. Schwartz, and T. M. Jessell, *Principles of Neuroscience*, 4th edition, p. 617, McGraw-Hill, 2000.

10. Faint noise boosts sensitivity to a sinusoidal signal in patients with a Nucleus-22 cochlear implant (from Advanced Bionics Corporation): "Introducing uniformly distributed, pseudorandom noise into the carrier envelope produced level-dependent effects. . . . At less sensitive low carrier levels, modulation sensitivity showed a stochastic resonance (SR) signature with increasing noise, displaying maximum sensitivity at an optimal noise level." M. Chatterjee and M. E. Robert, "Stochastic Resonance in Temporal Processing by Cochlear Implant Listeners?" *Fluctuations and Noise in Biological, Biophysical, and Biomedical Systems*, S. M. Bezrukov, H. Frauenfelder, and F. Moss, eds., *Proceedings of SPIE*, vol. 5110, pp. 348–355, 2003. Related work on noise-enhanced cochlear models includes N. G. Stocks, D. Allingham, and R. P. Morse, "The Application of Suprathreshold Stochastic Resonance to Cochlear Implant Coding," *Journal of Noise and Fluctuation Letters*, vol. 2, pp. 169–181, 2002; M. Bennett, K. Wiesenfeld, and F. Jaramillo, "Stochastic Resonance in Hair Cell Mechanoelectrical Transduction," *Journal of Fluctuation and Noise Letters*, vol. 4, no. 1, pp. 1–10, 2004.

11. The cell phone study at the Karolinska Institute in Stockholm used data on Swedes with acoustical neuromas. It looked at 148 cases involving cell phone use against 604 randomly selected control cases and found a fourfold increase in the odds ratio: "The overall odds ratio for acoustic neuromas associated with regular

mobile phone use was 1.0 (95% confidence interval = 0.6 – 1.5). Ten years after the start of mobile phone use the estimated relative risk increased to 1.9 (0.9 – 4.1); when restricting to tumors on the same side of the head as the phone was normally used, the relative risk was 3.9 (1.6 – 9.5)." S. Lonn, A. Ahlbom, P. Hall, and M. Feychting, "Mobile Phone Use and the Risk of Acoustic Neuroma," *Epidemiology*, vol. 15, no. 6, pp. 653–659, November 2004.

12. The airport noise study looked at children who went from quiet environments to noisy environments and at those who went in the other direction from noisy to quiet environments: S. Hygge, G. W. Evans, and M. Bullinger, "A Prospective Study of Some Effects of Aircraft Noise on Cognitive Performance in School Children," *Psychological Science*, vol. 13, no. 5, pp. 469–474, September 2002.

13. *Newsweek* reports that Americans now own about 90 million leaf blowers and that the federal government no longer directly funds noise abatement as it once did: "In 1972, the Office of Noise Abatement and Control (ONAC) was created to identify sources of noise and combat them. But in 1981, Congress and the Reagan Administration eliminated ONAC funding." J. Kluger, "Just Too Loud," *Newsweek*, p. 55, 5 April 2004. The text quote is from page 56.

14. Florida manatees can weigh up to 1,300 kilograms. Some have survived more than a dozen encounters with a boat propeller. Dedicated signaling hardware may be a feasible way to allow them to coexist with boats: "In light of . . . the known acoustical characteristics of shallow-water habitats, the spectra of boat noise and the dangerous, deceptive problem of acoustical shadowing, it is apparent that manatees, and perhaps other passive-listening marine mammals, could benefit from an acoustic warning device designed to fit on the front of boats, ships, and barges." E. R. Gerstein,

"Manatees, Bioacoustics, and Boats," *American Scientist*, vol. 90, pp. 154–163, March 2002.

15. There is circumstantial evidence of sonar-induced "barotrauma" in beaked whales: "Since the 1970s there have been at least three reports of unusual strandings of beaked whales in close proximity to military exercises. . . . In the Bahamas researchers noted that many of the stranded whales were bleeding from the ears. And dissections and computerized tomography scans of the salvaged tissues . . . found hemorrhaging and other telltale signs of barotraumas." D. Malakoff, "A Roaring Debate Over Noise," *Science*, vol. 291, pp. 576–578, 26 January 2001.

16. The study is controversial because of the circumstantial link between navy sonar and apparent decompression sickness in whales: "Here we present evidence of acute and chronic tissue damage in stranded cetaceans [beaked whales] that results from the formation *in vivo* of gas bubbles, challenging the view that these mammals do not suffer decompression sickness. The incidence of such cases during a naval sonar exercise indicates that acoustic factors could be important in the etiology of bubble-related disease." P. D. Jepson et al., "Gas-Bubble Lesions in Stranded Cetaceans," *Nature*, vol. 425, pp. 575–576, 9 October 2003. Jepson and his colleagues replied to critics and made clear that there is no conclusive causal link between sonar and the bends: "We stated neither that DCS [decompression sickness] occurs naturally in cetaceans, nor that exposure to active sonar increases its occurrence." A. Fernandez et al., "Pathology: Whales, Sonar, and Decompression Sickness (Reply)," *Nature*, doi: 10.1038/nature02528, 15 April 2004.

17. The researchers used a calibrated hydrophone array to measure the song lengths of 16 male humpback whales: "At least two songs

were recorded before the observation vessel requested the U.S. Navy R/V *Cory Chouest* to transmit ten (in one case four) 42-second LFA (low-frequency-active sonar) signals at 6-minute intervals. The sonar was broadcast at less than full strength, and no focal singer was exposed to a signal louder than 150 decibels (with respect to $1\mu Pa$)." P. J. O. Miller, N. Biassoni, A. Samuels, and P. L. Tyack, "Whale Songs Lengthen in Response to Sonar," *Nature*, vol. 405, p. 903, 22 June 2000.

18. Great tits showed song plasticity in noise environments that ranged from 42 to 63 decibels and that included highway noise: "We compared noise amplitude with the spectral distribution of sound energy within the range of the minimum frequency of great-tit song and found that in noisy territories there is a greater proportion of sound energy in the lower half of this range than in quiet territories (Pearson's $r = .78$, p-value $< .001$)." H. Slabbekoorn and M. Peet, "Birds Sing at a Higher Pitch in Urban Noise," *Nature*, vol. 242, p. 267, 17 July 2003.

19. For decibel-based noise maps of Europe see for instance the Web site http://www.xs4all.nl/~rigolett/ENGELS/maps/euromapclick.htm. For Paris noise maps (in French) see http://www.v1.paris.fr/fr/Environnement/bruit/. For the noise policy of the European Union see http://europa.eu.int/comm/environment/noise/home.htm#2.

CHAPTER 4: WHITE NOISE AIN'T SO WHITE

1. Brownian motion is a peculiar random process. It is continuous with probability one and yet it has so many "kinks" in its time samples that it is nondifferentiable with probability one. It is also self-similar: Zoom into a section of Brownian noise and it still

looks like jagged Brownian motion. Zoom into that smaller section and it too looks like and is jagged Brownian motion. Continue zooming in and the result will still look like and be jagged Brownian motion. Still Brownian motion B has a mean-square derivative process that equals Gaussian white noise n because Brownian motion has the autocorrelation function $r_B(t, s) = \min(t, s)$ and so its time-derivative process has the autocorrelation function $r_{dB}(t, s) = \sigma^2 \delta(s) = r_n(s)$ and because Brownian motion also has independent increments and Gaussian marginal probability density functions.

A noise process $\{n(t)\}$ is Gaussian white noise if all its samples are jointly Gaussian and if it is independent in time. We further assume that it is wide-sense stationary (WSS) so that its first and second moments are invariant to time shifts—this gives strict-sense stationarity for a Gaussian process. Uncorrelated Gaussian random variables are independent and so in this case independence just means that the noise process has a delta-pulse autocorrelation function: $r_{nn}(s) = E[n(t)n^*(t-s)] = \sigma^2 \delta(s)$ where the asterisk denotes complex conjugation. Typical signal processing notation for a *discrete-time* WSS zero-mean white Gaussian noise process is $w[n] \sim WGN(0, \sigma_w^2)$ as in D. G. Manolakis, V. K. Ingle, and S. M. Kogon, *Statistical and Adaptive Signal Processing*, p. 110, Prentice Hall, 2000. So it has a flat power spectrum $R_{ww}(e^{i\omega}) = \sigma_w^2$ across any frequency interval of length 2π because discrete-time power spectra are 2π-periodic.

The formal definition of the Brownian motion process $\{B(t)\}$ for $t \geq 0$ requires that $B(0) = 0$ and that $B(t)$ is continuous at the origin 0. The increments $B(t+s) - B(s)$ are independent of all other such increments for all choices of the time indices and they are Gaussian: $B(t+s) - B(s) \sim \mathcal{N}(0, \sigma^2 t)$. Then the variance of $B(t)$ will have the time-varying form $\sigma^2 t$. Normalizing by the constant variance parameter σ^2 produces *standard* or standardized Brownian motion because the process has zero mean. This is a martingale process. The absolute-value process $|B(t)|$ is a *reflected* Brownian motion or rather reflected at the origin. See S. Karlin and H. M. Taylor, *A First Course in Stochastic Processes*, 2nd ed., pp. 340–352, Academic Press, 1975.

2. Sustained white noise can damage the auditory cortex of baby rats and perhaps of baby humans: "Rearing infant rat pups in continuous, moderate-level noise delayed the emergence of adult-like topographic representational order and the refinement of response selectivity in the primary auditory cortex (A1). . . . These results demonstrate that A1 [auditory cortex] organization is shaped by a young animal's exposure to salient, structured acoustic inputs—and implicate noise as a risk factor for abnormal child development." E. F. Chang, and M. M. Merzenich, *Science*, vol. 300, pp. 498–502, 18 April 2003.

3. The autumn colors of many tree and plant leaves tend to appear as the green chlorophyll molecules disappear: "Chlorophyll catabolism unmasks foliar carotenoids, which diminish or accumulate during senescence, depending on the plant species. . . . [T]he foliage of deciduous trees and many other plants develops vivid colors before being shed. Fruits also become brightly colored during ripening. In such cases, the loss of chlorophyll unmasks underlying carotenoids, which provide a yellow or orange background against which new pigments accumulate." B. B. Buchanan, W. Gruissem, and R. L. Jones, *Biochemistry & Molecular Biology of Plants*, p. 1059, American Society of Plant Physiologists, 2000.

4. Laws of large numbers (LLNs) state that the sample mean $\bar{X}_n$ of (finite-variance) independent and identically distributed (i.i.d.) random variables $X_1, X_2, \ldots$ converges to the population mean m with all types of stochastic convergence: $\bar{X}_n \to m$ where $\bar{X}_n = \frac{1}{n}\sum_{k=1}^{n} X_k$

and where $m = E[X_k]$ for all k. Convergence with probability one gives the *strong* LLN: $P(\lim_{n\to\infty} \bar{X}_n = m) = 1$. Mean-square convergence gives the *mean-square* LLN: $\lim_{n\to\infty} E[(\bar{X}_n - m)^2] = 0$. Both of these LLNs imply the *weak* LLN where the sample mean converges in probability to the population mean: For all $\varepsilon > 0$: $\lim_{n\to\infty} P(|\bar{X}_n - m| > \varepsilon) = 0$. All three LLNs in turn imply that the sample mean converges to the population mean in distribution or in the very weak sense that the cumulative distribution function of the sample mean converges pointwise at all points of continuity to the constant population mean. (This latter and weakest type of stochastic convergence is the convergence type of the central limit theorem discussed below.) Such averaging benefits i.i.d. *finite*-variance random variables because $V[\bar{X}_n] = \frac{\sigma^2}{n}$. So the variance of the sample mean decreases linearly with sample size or equivalently the standard error falls with the square root of n. But this benefit from averaging completely fails if the random variables are i.i.d. standard Cauchy (and thus have infinite variance) with probability density function $f(z) = \frac{1}{\pi(1 + z^2)}$. Then Cauchy's residue theorem gives the characteristic function ϕ_z or Fourier transform of the probability density function as $\phi_z(\omega) = e^{-|\omega|}$. But the characteristic function of a sum of independent random variables is just the product of the characteristic functions. Combining this fact with the identical standard Cauchy distribution of each random variable Z_n implies that the sample mean $\bar{Z}_n$ is also standard Cauchy because its characteristic function is standard Cauchy: $\phi_{\bar{Z}_n} = \phi_z$. Hence there are *no* benefits to averaging: The sample mean of a random sample of a trillion Cauchy random variables performs no better than picking at random any one of the trillion Cauchy random variables and using it instead as a predictor.

5. The central limit theorem (CLT) states that *standardized* sums Z_n of (finite-variance) independent and identically distributed random

variables $X_1, X_2, \ldots$ converge in distribution to a standard normal random variable Z: $Z_n = \dfrac{\bar{X}_n - m}{\sigma/\sqrt{n}} \to Z \sim \mathcal{N}(0, 1)$ as the sample size n increases to infinity for finite mean m and finite standard deviation σ and for the sample-mean random variable $\bar{X}_n = \frac{1}{n} \sum_{k=1}^{n} X_k$. Convergence in distribution means that the cumulative distribution function of Z_n converges pointwise to the standard normal cumulative distribution function at all points x of continuity: $\lim_{n \to \infty} F_{Z_n}(x) = \frac{1}{\sqrt{2\pi}} \int_{-\infty}^{x} e^{-z^2/2} dz$. Another way to see the crucial role of standardization is to rewrite Z_n as $Z_n = \dfrac{1}{\sqrt{n}} \sum_{k=1}^{n} \left(\dfrac{X_n - m}{\sigma} \right)$. So the key normalization term is the *square root* of the sample size n and not (as in laws of large numbers) the sample size n itself. Hence either way $\bar{X}_n \approx \mathcal{N}(m, \sigma^2/n)$ for "large" n—typically for a sample size of at least $n > 30$. A common error is to conclude from the CLT that $\bar{X}_n \approx \mathcal{N}(m, \sigma^2)$. This ignores the square-root normalizer and wrongly concludes that the sample mean converges to a random variable rather than to a constant (the population mean). Hence such reasoning would contradict the laws of large numbers.

6. It takes at least 1,068 random samples to produce a 95% confidence interval of the unknown population proportion p with a margin of error of 3%. It takes at least 1,849 random samples to produce a 99% confidence interval with the same 3% margin of error. This arises from first applying the central limit theorem to the estimate $\hat{p}$ of the random success proportion $Y = (X_1 + \cdots + X_n)/n$ for n independent binary Bernoulli random variables X_k (each a model of the voter or poll subject who answers either yes or no to a question) with unknown proportion or success probability p. Then the maximum likelihood estimator $\hat{p}$ is just the sample mean of poll responses: $\hat{p} = \bar{X}_n$. The central limit theorem applies for sufficiently large sample sizes n and implies that $\hat{p}$ is approximately normal: $\hat{p} \sim \mathcal{N}(p, p(1-p)/n)$. This in turn gives the desired $(1 - \alpha)\%$

confidence interval as $\hat{p} \pm z_{\alpha/2} \sqrt{\hat{p}(1-\hat{p})/n}$ for the standard normal z-value $z_{\alpha/2}$ because the confidence interval is two-sided. Let e denote the margin of error in the estimate of p: $e = z_{\alpha/2} \sqrt{\hat{p}(1-\hat{p})/n}$. Then a conservative estimate of the required sample size for a given margin of error is $n = z_{\alpha/2}^2 / 4e^2$. For details of the argument see R. V. Hogg and E. A. Tanis, *Probability and Statistical Inference*, 7th ed., p. 380, Prentice Hall, 2006. A 95% confidence interval corresponds to a two-sided z-value of 1.96. Thus a required 3% margin of error leads to $n = (1.96)^2 / 4(.03)^2 = 1067.1111$ or at least 1,068 samples. A 99% confidence interval corresponds to a z-value of 2.58. It requires $n = (2.58)^2 / 4(.03)^2 = 1,849$ samples. This 73% increase in sample size explains why media polls tend to use the 95% confidence interval rather than the more accurate but much more expensive 99% confidence interval.

7. Stable probability density functions are a subclass of infinitely divisible distributions. Stable densities have no known closed form except in the symmetric case for the Gaussian and Cauchy densities and in the asymmetric case for the Levy density. So researchers must compute most stable densities by taking the Fourier transform of the corresponding characteristic function ϕ. The stable characteristic functions have the form $\phi(\omega) = \exp\{ia\omega - \gamma|\omega|^\alpha (1 + i\beta \, \mathrm{sgn}(\omega) \, \tan(\alpha\pi/2))\}$ if $\alpha \neq 1$ and otherwise $\phi(\omega) = \exp\{ia\omega - \gamma|\omega| (1 - 2i\beta \ln \, \mathrm{sgn}(\omega)/\pi)\}$ if $\alpha = 1$ where i is the imaginary unit $i = \sqrt{-1}$ The location parameter a acts as the "mean" or center of the probability density function when $a > 0$. The skewness parameter β must be zero for a stable bell curve and in general obeys $-1 \leq \beta \leq 1$. Such symmetric alpha-stable or $S\alpha S$ bell curves are symmetric about a if $\beta = 0$. The dispersion parameter γ acts like a variance (which exists only in the Gaussian case) because it controls the width of a $S\alpha S$ bell curve. Stochastic resonance or noise-benefit plots show a system performance measure

such as mutual information or a signal-to-noise ratio as a function of this stable noise dispersion parameter. Note that the characteristic functions of zero-centered $S\alpha S$ probability density functions have the substantially simpler form $\phi(\omega) = \exp\{-\gamma|\omega|^\alpha\}$ since $a = \beta = 0$.

The parameter α controls the tail thickness of an $S\alpha S$ bell curve and in general obeys $0 < \alpha \le 2$. The tail becomes thicker as α decreases in value. The variance of a stable density does not exist if $\alpha < 2$ even though *fractional* lower-order moments do exist in this general case. The special $S\alpha S$ case $\alpha = 2$ or $\phi(\omega) = \exp\{-\gamma\omega^2\}$ corresponds to the Gaussian or normal probability density. The Gaussian alone has exponentially decaying tails while other $S\alpha S$ bell curves have tails that decay more slowly as power laws. The still simpler characteristic function $\phi(\omega) = \exp\{-\omega^2\}$ defines a zero-mean Gaussian random variable with variance $\sigma^2 = 2$. The standard Cauchy probability density function $f(x) = \dfrac{1}{\pi(1 + x^2)}$ has quite thick tails and corresponds to the simplest stable characteristic function of all: $\phi(\omega) = \exp\{-|\omega|\}$. A Levy random variable with asymmetric stable density corresponds to $\alpha = \frac{1}{2}$ and $\beta = 1$. Stable random variables do not in general have covariances but they do have a more complex measure of joint dispersion called covariations. These covariations generalize the hyperellipsoids of multidimensional Gaussian and other finite-variance densities and can form the basis of clustering algorithms and fuzzy-rule generators. For details see H. M. Kim and B. Kosko, "Fuzzy Prediction and Filtering in Impulsive Noise," *Fuzzy Sets and Systems*, vol. 77, no. 1, pp. 15–33, January 1996. The term "stable" refers to the fact that sums of stable random variables (all with the *same* parameter α) are again stable. This generalizes the well-known property that sums of Gaussian random variables are again Gaussian.

For further information on stable variables and free stable software see the online Web page of Professor John Nolan at http://academic2.american.edu/~jpnolan/stable/stable.html. Nolan's site and related materials contain a proof of the *generalized central limit theorem*: Suppose $X_1, X_2, \ldots$ are independent and identically distributed random variables. Then there exist positive constants a_n and real constants b_n and a nondegenerate random variable Z such that the transformed sum $a_n \left(\sum_{k=1}^{n} X_k \right) - b_n$ converges in distribution to Z if and only if Z is a stable random variable for some α in $(0, 2]$. The classical central limit theorem is the special case when Z is the standard normal random variable and when the random variables X_k have finite variance σ^2 and finite mean m—and when $a_n = 1/(\sigma\sqrt{n})$ and $b_n = m\sqrt{n}/\sigma$.

8. The "bursty" nature of Ethernet traffic collected at Bellcore Laboratories appears to best fit an asymmetric stable model with stable parameter $\alpha = 1.7$: "The proposed [stable and self-similar] model accepts physical interpretation since it explains how the observed data appear as a superposition of independent effects. The reason is that its underlying marginal distribution is alpha-stable and therefore satisfies the generalized central limit theorem." A. Karasaridis and D. Hatzinakos, "Network Heavy Traffic Modeling Using Alpha-Stable Self-Similar Processes," *IEEE Transactions on Communications,* vol. 49, no. 7, pp. 1203–1214, July 2001. For a mathematical analysis of related modeling issues see T. Mikosch, S. Resnick, H. Rootzen, and A. Stegeman, "Is Network Traffic Approximated by Stable Levy Motion or Fractional Brownian Motion," *Annals of Probability,* vol. 12, no. 1, pp. 23–68, 2002. A different approach is to find the best-fitting P-P or probability-percentile plot for a data set but allowing stable models as viable candidates. One such study found that the best-fitting bell curve included a symmetric alpha-stable probability density with tail-thickness parameter $\alpha = 1.228$ and thus with a near-Cauchy level of impulsiveness: A. Briassouli, P. Tsakalides, and A. Stouraitis,

"Hidden Messages in Heavy Tails: DCT-Domain Watermark Detection Using Alpha-Stable Models," *IEEE Transactions on Multimedia,* vol. 7, no. 4, pp. 700–715, August 2005.

9. Uncertainty principles arise from the Cauchy-Schwartz inequality and state that a product of two standard deviations exceeds some positive constant: $\sigma_X \sigma_Y > c > 0$ for random variables X and Y. This hyperbolic inequality implies that decreasing the variance-measured "uncertainty" of Y must increase the corresponding uncertainty of X because then $\sigma_X > c/\sigma_Y$. That uncertainty tradeoff would not follow if the two random variables were uncorrelated and thus had zero covariance $\alpha_{XY} = E[(X - E(X))(Y - E(Y))]$. The general (classical) uncertainty principle for finite-variance random variables is $\sigma_{XY}^2 \leq \sigma_X^2 \, \sigma_Y^2$ and thus does not depend on physics or chemistry or anything other than the assumption of finite variances (which classical quantum mechanics assumes in its Hilbert-space structure). Signal processing uses the special case involving the variance in a time signal $s(t)$ and the variance of its corresponding (normalized) Fourier transform $S(\omega)$: $1/4 \leq \sigma_s^2 \sigma_S^2$ with equality just in case the time signal is Gaussian (A. Boggess and F. Narcowich, *A First Course in Wavelets with Fourier Analysis,* pp. 120–125, Prentice Hall, 2001). Quantum (nonclassical) uncertainty principles result from linear operators that do not commute as in the famous Heisenberg inequality for the standard deviation of position x and of the corresponding momentum p: $\hbar/2 \leq \sigma_X \sigma_p$ where h is Planck's constant 6.6×10^{-34} joules seconds and where $\hbar = h/2\pi$. For a simple derivation see D. C. Marinescu and G. M. Marinescu, *Approaching Quantum Computing,* pp. 87–89, Prentice Hall, 2005.

10. Chaos and noise often resemble each other and leave similar footprints. It is hard to distinguish between a bona fide chaotic pattern and some form of colored noise based only on time-series

measurements from the olfactory bulb of a rabbit: "On the basis of further modeling with nonlinear differential equations and a review of our data reflecting experimentally induced sustained oscillatory states of the olfactory system, we now believe that the EEG 'burst' that carries the perceptual information from the bulb to the prepyriform cortex is not the manifestation of a limit cycle but reflects instead another chaotic attractor. We do not offer this as a firm conclusion, but rather as a more plausible hypothesis than the one originally presented. It is likely that there will remain substantial uncertainty for some years, possibly decades, about the differences between a limit cycle trajectory that aborts prior to convergence to a periodic attractor, a limit cycle attractor under perturbation by noise, and a narrow spectral band chaotic attractor in which the unpredictability appears in variation of phase or in frequency narrowly about a mean." W. J. Freeman and C. A. Skarda, "Chaotic Dynamics versus Representationalism," *Behavioral and Brain Sciences*, vol. 13, no. 1, pp. 167–168, 1990. See also W. J. Freeman, "Random Activity at the Microscopic Neural Level in Cortex ('Noise') Sustains and Is Regulated by Low-Dimensional Dynamics of Macroscopic Cortical Activity ('Chaos')," *International Journal of Neural Systems*, vol. 7, no. 4, pp. 473–480, 1996.

11. Fuzzy or vague sets are sets whose elements belong to the set to some degree between 0 or 1 and hence they generalize classical bivalent sets. The adjective "fuzzy" stems from the landmark 1965 paper by Lotfi Zadeh of UC Berkeley: L. A. Zadeh, "Fuzzy Sets," *Information and Control*, vol. 8, pp. 338–353, 1965; also in L. A. Zadeh, *Fuzzy Sets and Applications: Selected Papers*, ed. R. R. Yager et al., New York: Wiley, 1987. For an alternative geometric view of (finite) fuzzy sets as points in unit hypercubes see the textbooks B. Kosko, *Neural Networks and Fuzzy Systems: A Dynamical Systems Approach*

to *Machine Intelligence*, Prentice Hall, 1992; B. Kosko, *Fuzzy Engineering*, Prentice Hall, 1996.

A formal definition of a fuzzy set assumes an arbitrary space or ground set X. Then $A \subset X$ is a fuzzy or multivalued subset if and only if A has the *set function* $a: X \to [0, 1]$. A is a binary set if and only if the set function a maps to exactly the two extremal values of 0 and 1: $a: X \to \{0, 1\}$. Set A is properly fuzzy if and only if both $A \cap A^c \neq \varnothing$ and $A \cup A^c \neq X$ hold and thus just in case it violates the binary "laws" of noncontradiction and excluded middle. We define intersection and union pointwise with minimum $a \cap a^c(x) = \min(a(x), 1 - a(x))$ and maximum $a \cap a^c(x) = \max(a(x), 1 - a(x))$ and define set complement with order reversal $a^c(x) = 1 - a(x)$. If the space X has n elements in the finite case or $X = \{x_1, \ldots, x_n\}$ then each of the uncountably many fuzzy sets A corresponds to a "fit" or *fuzzy unit* vector in the n-dimensional unit hypercube: $a = (a_1, \ldots, a_n) \in [0, 1]^n$ with pointwise versions in the more general continuous cases. These fuzzy cubes also offer a way to study and visualize the behavior of probabilistic conditioning operators: B. Kosko, "Probable Equality, Superpower Sets, and Superconditionals," *International Journal of Intelligent Systems*, vol. 19, pp. 1151–1171, December 2004. Many basic operations on fuzzy sets use some form of fuzzy cardinality or size $c(A)$ of a fuzzy set where $A \subset X$ has the fuzzy cardinality or count $c(A) = \sum_{i=1}^{n} a_i$ or $c(A) = \sum_{i=1}^{\infty} a_i < \infty$ for a finite fuzzy set $A = (a_1, \ldots, a_n) \in [0, 1]^n$ or a summable denumerable set $A = (a_1, a_2, \ldots) \in [0, 1]^{\infty}$ or where $c(A) = \int_{R^n} a(x)\,dx < \infty$ for a real fuzzy subset $A \subset R^n$ with arbitrary integrable joint set function $a: R^n \to [0, 1]$. Then the fuzzy equality operator E measures how similar or how equal two fuzzy sets or patterns A and B are to each other: $E(A, B) = Degree(A = B) = \dfrac{c(A \cap B)}{c(A \cup B)}$. The special case of $B = A^c$ gives back a key measure of fuzziness or vagueness of the fuzzy set A: $Fuzziness(A) = E(A, A^c) = \dfrac{c(A \cap A^c)}{c(A \cup A^c)}$ or a ratio of the counted violations of the "laws" of noncontradiction and excluded middle. Fuzzy sub-sethood or degree of set inclusion $S(A, B) = Degree(A \subset B) = \dfrac{c(A \cap B)}{c(A)}$ corresponds to how much A resembles $A \cap B$ since $S(A, B) = E(A, A \cap B)$. The subsethood operator can likewise eliminate the fuzzy equality operator. All three operators coincide in the special non-binary case where A relates to its opposite A^c: $Fuzziness(A) = S(A \cup A^c, A \cap A^c) = E(A, A^c)$. Note that the probabilistic relative frequency or ratio of counted successes to trials $\dfrac{n_A}{n}$ corresponds to $S(X, A) = \dfrac{c(X \cap A)}{c(X)} = \dfrac{c(A)}{n} = \dfrac{n_A}{n}$. So the purely "random" concept of relative frequency is here just the fuzzy "whole in the part" or the partial inclusion of the sample space X in the event A—a relation that cannot occur if the subsethood operator is binary.

Fuzzy systems use sets of fuzzy if-then rules to map inputs to outputs and so they define functions or mappings $F:R^n \to R^p$ that convert an input n-vector x into an output p-vector $F(x)$. Most fuzzy systems are some form of an additive fuzzy system. These systems add together partially fired rules and then compute the final output by taking the centroid of the added partially fired then-parts of the rules. This in turn shows that such fuzzy systems compute a conditional expectation $F(x) = E[Y|X=x]$ with a corresponding second-order conditional variance or covariance. Feedback fuzzy systems also exist but stability conditions so far impose extremely limiting constraints on the structure of the fuzzy system's then-part rules. For details see B. Kosko, *Fuzzy Engineering*, Prentice Hall, 1996. The system conditional variances give a measure of the intrinsic uncertainty of any fuzzy-system answer $F(x)$ to a given input or question x since the fuzzy system computes answers by interpolating its rule structure. This conditional variance has a simple form in the scalar case when all then-part sets have the same shape. For an application of this second-order uncertainty to modeling gunshot bruises see I. Y. Lee, B. Kosko, and

W. F. Anderson, "Modeling of Gunshot Bruises in Soft Body Armor with an Adaptive Fuzzy System," *IEEE Transactions on Systems, Man, and Cybernetics*, vol. 35, no. 6, pp. 1374–1390, December 2005.

The fuzzy approximation theorem (FAT) says that an additive fuzzy system $F:C \subset R^n \to R^p$ with a finite number m of rules can uniformly approximate any continuous or bounded measurable function $f:C \subset R^n \to R^p$ if the domain C is a compact (closed and bounded) set. In this sense fuzzy systems are dense in the space of continuous (or bounded measurable) functions much as the rational numbers are dense in the real numbers. The if-part fuzzy sets $A_j \subset R^n$ can have any multivalued set function $a_j:R^n \to [0, 1]$ and so can have any shape. The then-part sets $B_j \subset R^p$ can also be arbitrary so long as the then-part set functions $b_j: R^p \to [0, 1]$ are integrable. The proof of the FAT theorem is constructive and uses a fuzzy cover of rule patches $A_j \times B_j \subset R^n \times R^p$ with a product Cartesian set function $R_{Aj \to Bj}(x, y) = a_j(x) b_j(y)$ for the jth rule of verbal form "If X is A_j then Y is B_j" or $A_j \to B_j$. This constructive graph cover in theory allows clustering techniques and other learning schemes to learn the rules given enough time and enough accurate input-output samples from the approximand function f. For details see B. Kosko, "Fuzzy Systems as Universal Approximators," *IEEE Transactions on Computers*, vol. 43, no. 11, pp. 1329–1333, November 1996; an earlier version appears in the *Proceedings of the First IEEE International Conference on Fuzzy Systems (FUZZ-92)*, pp. 1153–1162, March 1992. There is in general no best shape for the if-part fuzzy sets just as there is no general best shape of a probability density function. Some set shapes do perform better than others in the task of adaptively approximating given test functions: S. Mitaim and B. Kosko, "The Shape of Fuzzy Sets in Adaptive Function Approximation," *IEEE Transactions on Fuzzy Systems*, vol. 9, no. 4, pp. 637–656, August 2001. Often a good choice for an if-part set function is some scaled form of the "sinc" function $\sin(x)/x$ that also appears in the sampling theorem of signal processing.

12. So far we can solve the Schrödinger wave equation $(i\hbar\frac{\partial\psi}{\partial t}=H\psi)$ only for "hydrogenlike" atoms that have only one surrounding electron. That lets us avoid the many-body problem or "curse of dimensionality" that results if we instead take account of the multiple interacting electron shells that surround the nucleus of non-hydrogen atoms: "Instead of treating just the hydrogen atom, we consider a slightly more general problem: the *hydrogenlike* atom. By this we mean a system consisting of one electron and a nucleus of charge Ze. For $Z=1$, we have the hydrogen atom. For $Z=2$, the He$^+$ (helium) ion. For $Z=3$, the Li^{++} (lithium) ion, etc. The hydrogenlike atom is the single most important system in quantum chemistry. An exact solution of the Schroedinger equation for atoms with more than one electron cannot be obtained due to the interelectronic repulsions." I. N. Levine, *Quantum Chemistry*, 2nd ed., p. 98, Allyn and Bacon, 1974.

13. Crackle noise appears to occur in many applications but the phenomenon itself still lacks a clear explanation: "But the successes [of partial theoretical explanations] remain dwarfed by the bewildering variety of systems that crackle. Achieving a global perspective on the universality classes for crackling noise remains an open challenge." J. P. Sethna, K. A. Dahmen, and C. R. Myers, "Crackling Noise," *Nature*, vol. 410, pp. 242–250, 8 March 2001.

14. The search for pink noise involves a great deal of controversy over whether it occurs at all and which nonlinear mechanisms might produce it if it in fact occurs. The strongest evidentiary and theoretical case for pink noise remains at the level of electrical circuits. For a good review see F. N. Hooge, "1/f Noise Sources," *IEEE Transactions* on *Electron Devices*, vol. 41, no. 11, pp. 1926–1935, November 1994. For a related critique of alleged findings of fractal behavior see D. Avnir, B. Ofer, D. Lidar, and O. Malcai, "Is the Geometry of Nature Fractal?" *Science*, vol. 279, pp. 39–40, 2 January 1998 (and related replies and responses). For a mathematical review of pink-noise mechanisms and controversies see E. Milotti, "1/f Noise: A Pedagogical Review," *Archives of Physics*, 0204033, pp. 1–26, 2002.

15. There is some theoretical evidence for the persistence of black noise relative to pink noise in geophysical data based on studies of autoregressive (all-pole) statistical models: "We conclude that populations exposed to red noise should be considered more 'at risk' than populations exposed to black noise and that more conservative management strategies should be adopted." K. M. Cuddington and P. Yodzis, "Black Noise and Population Persistence," *Proceedings of the Royal Society of London–B*, vol. 266, pp. 969–973, 1999.

16. Classical thermodynamics states that the scaled absolute temperature T of a molecular gas equals the average kinetic energy of the gas molecules: $\frac{1}{2}m\overline{v^2}=\frac{3}{2}kT$ where m is the mass of any one of the identical molecules that has an ensemble-averaged mean-square velocity $\overline{v^2}$. The Boltzmann constant k is 1.38×10^{-23} joules per kelvin. The absolute temperature T is in degrees kelvin such that absolute zero corresponds to $-273.15°C$ or $-459.67°F$. See D. H. Trevena, *Statistical Mechanics*, p. 48, Horwood Publishing Limited, 2001.

17. Many types of sky noise interfere with radar signal detection: "The principal source of antenna noise is sky noise. Sky noise consists of cosmic noise, which predominates at the lower frequencies, and atmospheric noise, which predominates at the higher frequencies. Maximum cosmic noise occurs in the direction of the galactic center and is minimum at the galactic pole.

Maximum atmospheric noise occurs with the antenna beam pointed along the horizon and minimum with the antenna pointed at the zenith." J. D. DiFranco, and W. L. Rubin, *Radar Detection*, pp. 459–460, Artech House, Inc., 1980.

18. The heat equation in one dimension is the classical model of diffusion or how changes in time depend on changes in space: $\frac{\partial p}{\partial t} = D \frac{\partial^2 p}{\partial x^2}$ for some functional p of the system variable x and some diffusion constant D. Einstein arrived at this diffusion equation through physical insight and an approximation argument in his famous 1905 paper on the mathematical structure of Brownian motion: A. Einstein, "Investigations on the Theory of the Brownian Movement," English translation, pp. 12–18, Dover Publications, 1956. Einstein viewed p as the probability $p(x(t) \mid x(0) = x_0)$ that the Brownian motion would be in location $x(t)$ at time t if it started from initial point x_0 at time zero. Einstein then observed that the time-varying Gaussian probability density function $p(x(t) \mid x_0) = (2\pi t)^{-1/2} \exp\{-(x - x_0)^2 / 2t\}$ satisfies the diffusion equation if we pick physical constants so that $D = \frac{1}{2}$: S. Karlin and H. M. Taylor, *A First Course in Stochastic Processes*, 2nd ed., p. 340, Academic Press, 1975. Einstein further observed that for a randomly wandering free particle the diffusion constant has the form $D = 2kT/f$ where k is Boltzmann's constant (the ratio of the gas constant to Avogadro's number), T is temperature in degrees kelvin, and f is the coefficient of friction. More general Brownian models in physics arise from a randomized version of Newton's second law of motion in which a random disturbance or collision (noise) term adds to a frictional term or other forcing term that balances the particle's mass times its acceleration. Physicists often call this Brownian description a Langevin equation and use some form of the Fokker-Planck equation to find its controlling and time-evolving probability density function. For the history and original source materials on this evolving family of Brownian models see N. Wax, ed., *Selected Papers on Noise and Stochastic Processes*, Dover Publications, 1954. For a modern and thorough review of Einstein's Brownian achievement and its legacy see the excellent review article commemorating the hundredth anniversary of Einstein's discovery: L. Cohen, "The History of Noise," *IEEE Signal Processing Magazine*, vol. 22, no. 6, pp. 20–45, November 2005.

19. Engineers often state Nyquist's theorem for thermal resistor noise in an electrical circuit as a simple variance approximation: $V_{rms}^2 = 4kTRB$. Here V_{rms} is the root-mean-square of the noise voltage in a Thevenin equivalent electrical circuit, $k = 1.38 \times 10^{-23}$ joules per kelvin is Boltzmann's constant, T is the absolute resistor temperature in degrees kelvin, R is the resistance in ohms, and B is the effective (one-sided) bandwidth. See J. R. Cogdel, *Modeling Random Systems*, pp. 608–612, Prentice Hall, 2004. Then the emitted power spectrum $S(f)$ in a matched load is approximately $S(f) = kT/2$ in watts per hertz (two-sided) and hence does not depend on the frequency f. So the noise spectrum $S(f)$ stays flat or whitelike until quantum rolloff occurs at about $k = 2.5 \times 10^{10} T$ hertz at roughly room temperatures.

Nyquist himself was careful to state the more general quantum version of his result at the very end of his article: "If the energy per degree of freedom be taken [as] $hv/(e^{hv/kT} - 1)$ where h is the Planck constant [$h = 6.6 \times 10^{-34}$ joules seconds], the expression for the electromotive force in the interval dv becomes $E_v^2 dv = 4Rhdv/(e^{hv/kT} - 1)$. Within the ranges of frequency and temperature where experimental information is available this expression is indistinguishable from that obtained from the equipartition law [nonquantum case]." H. Nyquist, "Thermal Agitation of Electrical Charge in Conductors," *Physical Review*, vol. 32, pp. 110–113 at 113, July 1928. A typical modern version gives the quantum-mechanical noise power spectral density $S_n(f)$ based on quantum radiation as the ratio $S(f) = hf/2(e^{hf/kT} - 1)$. The ratio defines a wide and roughly flat curve with a maximum at $f = 0$ (using L'Hôpital's rule for limits of indeterminates) of $kT/2$ or a noise power of $\mathcal{N}_0/2$ in the notation of communications engineering. This also follows for low frequencies from the first two terms of the power series of the exponential: $e^{hf/kT} \approx 1 + hf/kT$. The monotonic quantum rolloff is negligible over much of the spectrum. It is only about 90% of this maximum in the far infrared portion of the electromagnetic spectrum near one terahertz while at the room temperature of $T = 300°K$. The rolloff increases rapidly from then on and ensures a total finite noise power. The communications notation also explains the typical engineering notation for (wide-sense stationary and ergodic Gaussian) white noise in terms of the autocorrelation r_{XX} function as $r_{XX}(\tau) = \mathcal{N}_0 \, \delta(\tau)/2$ as in J. G. Proakis and M. Salehi, *Communication Systems Engineering*, p. 191, Prentice Hall, 1994.

20. Noise arises in the brain not only because of crosstalking neurons but also because of noisy electrical biophysical circuits and cellular components: "One major source of noise within neurons is voltage-gated ion channels embedded in the neuronal membrane. These channels are macromolecules which are subject to random changes of conformational state due to thermal agitation, and when these changes occur between a conducting and non-conducting state, the channel acts as a microscopic source of noise current which is injected into the cell. The noise current can change the spiking behavior of neurons, affecting the distribution of response latencies, spike propagation in branched cable structures, the generation of spontaneous action potentials, and the reliability and precision of spike timing [citations omitted]." P. N. Steinmetz, A. Manwani, C. Koch, M. London, and I. Segev, "Subthreshold Voltage Noise Due to Channel Fluctuations in Active

Neuronal Membranes," *Journal of Computational Neuroscience*, vol. 9, pp. 133–148, 2000. See also C. Koch, *Biophysics of Computation: Information Processing in Single Neurons*, pp. 194–211, 1999; W. Gerstner and W. Kistler, *Spiking Neuron Models: Single Neurons, Populations, Plasticity*, pp. 148–150, Cambridge University Press, 2002.

21. Tiny vibrating strings inside black holes may both bring about Hawking radiation and preserve the information structure of the matter that fell into the black hole: "This appears to be rather an extreme change in our picture of the black hole, particularly since (b) [which states that black hole 'hair' requires that the hole's microstates have no horizon or singularity] requires that the geometry of individual states differ significantly from the standard black hole metric everywhere in the interior of the hole, and not just within Planck distance of the singularity." S. D. Mathur, A. Saxena, and Y. Srivastava, "Constructing 'Hair' for the Three Charge Hole," *Nuclear Physics B*, vol. 680, pp. 415–449, March 2004.

22. Theoretical physicist Anthony Zee derives the temperature T of Hawking radiation by rotating the Schwarzchild metric for a black-hole solution and then making some physical approximations. This gives $T = \dfrac{\hbar c^3}{8\pi GM}$ where c is the speed of light, G is the gravitational constant, M is the mass of the black hole, and $\hbar = h/2\pi$ for Planck's constant h. See A. Zee, *Quantum Field Theory in a Nutshell*, p. 265, Princeton University Press, 2003. For the original source article see S. W. Hawking, "Particle Creation by Black Holes," *Communications in Mathematical Physics*, vol. 43, no. 3, pp. 199–220, 1975.

ThinkQuest's online library gives a simpler and largely classical derivation of the Hawking radiation temperature with the following computational examples. This leads to $T = \dfrac{hc^3}{8\pi kGM}$ with Boltzmann's constant k in the denominator. Then a black hole with the earth's mass of about 5.97×10^{11} kg would have a chilly temperature of only 0.02 kelvin or nearly absolute zero. The Stefan-Boltzmann law for the total energy flux of a radiating blackbody states that the flux is proportional to the fourth power of the temperature. This gives the black hole's luminosity as $L = \dfrac{32\pi^6 k^4 G^2 M^2 T^4}{15 h^3 c^6}$. Then a black hole with the earth's mass would have a trace luminosity of about 10^{-17} watts while a black hole of a hundred billion kilograms would have a luminosity of about 36 million kilowatts. The lifetime of a radiating (evaporating) black hole is about $t = M_{int}^3 / C$ years where M_{int} is the black hole's initial or starting mass and where the constant C is about 4×10^{15} kg^3/sec. Then the lifetime of the tiny black hole—one that is only a hundred billion kilograms—is about 3 billion years while the lifetime of a black hole with the mass of the sun (about 2×10^{30} kg) would be about 10^{67} years.

CHAPTER 5: FIGHTING NOISE WITH NOISE

1. Students in a first course on integral calculus may not know that the sinc function $f(x) = \dfrac{\sin x}{x}$ is one of the most important functions in modern signal processing. The sinc function reaches a maximum of unity at zero and falls off in decreasing undulations on either side as it crosses the x axis at integer multiples of π. The property that gives rise to the sampling theorem of signal processing is that suitably defined sinc functions are countably infinite wavelet bases that span the space of bandlimited functions. Sinc functions behave as binary functions for integer arguments: $\dfrac{\sin[\pi(n-m)]}{\pi(n-m)} = \begin{cases} 1 & \text{if } n=m \\ 0 & \text{if } n \neq m \end{cases}$. They have the further property that their Fourier transform defines a rectangle or ideal low-pass filter in the frequency domain. The ideal lowpass filter is the 2π-periodic

frequency response $H(e^{i\omega}) = \begin{cases} 1 & \text{if } |\omega| < \omega_c \\ 0 & \text{if } \omega_c < |\omega| \le \pi \end{cases}$ for cutoff frequency ω_c. Inverse-transforming gives the corresponding sinc function: $h[n] = \frac{1}{2\pi} \int_{-\omega_c}^{\omega_c} e^{i\omega n} d\omega = \frac{\sin \omega_c n}{\pi n}$. The first figure in chapter 5 illustrates this duality between the ideal low-pass filter in the frequency domain and the undulating sinc function in the time domain.

The sampling theorem uses a countably infinite set of samples and a countably infinite sum of interpolating sinc functions to exactly represent any bandlimited function. Suppose a continuous-time signal $x_c(t)$ is bandlimited to cutoff frequency Ω_c in the sense that its continuous-time Fourier transform $X_c(i\Omega)$ is zero outside the frequency interval $(-\Omega_c, \Omega_c)$: $X_c(i\Omega) = 0$ if $|\Omega| \ge \Omega_c$. Note 5 below defines the related discrete-time Fourier transforms. Define the uniformly spaced samples $x[n] = x_c(nT)$ for sampling period T to give the discrete-time sequence $x[n]$. Then the sampling theorem states that a convolution sum of interpolating sinc functions exactly represents the sampled continuous function: $\sum_{n=-\infty}^{\infty} x[n] \frac{\sin[\pi(t-nT)/T]}{\pi(t-nT)/T} = x_c(t)$ if $x_c(t)$ is bandlimited to Ω_c and if the sampling rate Ω_S is at least twice the *Nyquist rate* $2\Omega_c$–if $\Omega_S \equiv \frac{2\pi}{T} > 2\Omega_c$. For a proof see A. V. Oppenheim, R. W. Schafer, and J. R. Buck, *Discrete-Time Signal Processing*, 2nd ed., pp. 140–153, Prentice Hall, 1999.

2. The uncertainty principle of signal processing states that the product of the variance in a time signal $s(t)$ and the variance of its corresponding (normalized) Fourier transform $S(\omega)$ is bounded below by a positive constant: $\sigma_s^2 \sigma_S^2 \ge \frac{1}{4}$ with equality when the time signal is Gaussian. So increased precision in time implies decreased precision in frequency and vice versa. See note 9 in chapter 4 for a further discussion of such statistical inequalities based on the classical Cauchy-Schwartz inequality.

3. Statistical tests support the claim that some visual neurons in kitten brains minimize the time-frequency uncertainty principle: "Using two-dimensional (2-D) spatial and spectral response profiles described in the previous two reports, we test Daugman's generalization of Marcelja's hypothesis that simple receptive fields belong to a class of linear spatial filters analogous to those described by Gabor and referred to here as 2D Gabor filters. In the space domain we found 2D Gabor filters that fit the 2D spatial response profile of each simple cell in the least-squared error sense (with a simplex algorithm) and we show that the residual error is devoid of spatial structure and statistically indistinguishable from random error. . . . We conclude that the Gabor function provides a useful and reasonably accurate description of most spatial aspects of simple receptive fields. Thus it seems that an optimal strategy has evolved for sampling images simultaneously in the 2D spatial and spatial frequency domains." J. P. Jones and L. A. Palmer, "An Evaluation of the Two-Dimensional Gabor Filter Model of Simple Receptive Fields in Cat Striate Cortex," *Journal of Neurophysiology*, vol. 58, no. 6, pp. 1233–1258, 1987. See John Daugman's article for a mathematical description of 2D Gabor "logons": J. G. Daugman, "Uncertainty Relation for Resolution in Space, Spatial Frequency, and Orientation Optimized by Two-Dimensional Visual Cortical Filters," *Journal of the Optical Society of America A*, vol. 2, no. 7, pp. 1160–1169, July 1985.

4. Claude Shannon cited the mathematician J. M. Whittaker as the source of the sinc-based sampling theorem: "Theorem 1 [a version of the sinc-based sampling theorem] has been given previously in other forms by mathematicians (footnote 8) but in spite of its evident importance seems not to have appeared explicitly in the literature of communications theory." C. E. Shannon, "Communication in the Presence of Noise," *Proceedings of the Institute of*

Radio Engineers, vol. 37, pp. 10–21, 1949. Shannon's eighth footnote in the paper cites Whittaker as follows: "Whittaker, J. M., 'Interpolatory Function Theory,' *Cambridge Tracts in Mathematics and Mathematical Physics*, no. 33, Cambridge University Press, Chapt. IV, 1935."

5. A convolution is a time-reversed correlation. The discrete version is a sum of pairwise multiplications while the continuous version is an integral of pairwise multiplications. A discrete-time linear time-invariant (LTI) system has the form of a discrete convolution: $y[n] = x[n] * h[n] = \sum_{k=-\infty}^{\infty} x[k]h[n-k]$ for input sequence $x[n]$ and throughput or filter or "impulse response" sequence $h[n]$. Time-reversing the filter sequence gives the infinite sum its convolution form. This convolution in the time domain Fourier-transforms to multiplication in the frequency domain. Define the discrete-time Fourier transform (DTFT) of the input sequence $x[n]$ to be $X(e^{i\omega}) = \sum_{n=-\infty}^{\infty} x[n]e^{-i\omega n}$ and similarly for the DTFTs of the output and filter sequences. The DTFT is 2π-periodic and has inverse transform $x[n] = \frac{1}{2\pi}\int_{-\pi}^{\pi} X(e^{i\omega})e^{i\omega n}d\omega$. Then the convolution theorem converts the LTI convolution system into the product of DTFTs: $y[n] = x[n] * h[n] \leftrightarrow Y(e^{i\omega}) = X(e^{i\omega})H(e^{i\omega})$. Signal processors often call the 2π-periodic $H(e^{i\omega})$ the frequency response of the LTI system. For further details on such disrete-time signal processing see A. V. Oppenheim, R. W. Schafer, and J. R. Buck, *Discrete-Time Signal Processing*, 2nd ed., Prentice Hall, 1999.

6. The pearl in the oyster of random processing with LTI systems is the spectral relation $R_{yy}(e^{i\omega}) = R_{xx}(e^{i\omega})|H(e^{i\omega})|^2$. This result follows under fairly general conditions (rational power spectra) from passing a wide-sense stationary *random* input sequence $x[n]$ through the linear filter of an LTI system $y[n] = x[n] * h[n]$. The

power spectral density $R_{xx}(e^{i\omega})$ of the random sequence $x[n]$ is the DTFT of its correlation function: $R_{xx}(e^{i\omega}) = \sum_{k=-\infty}^{\infty} r_{xx}[k]e^{-i\omega k}$ for the autocorrelation function $r_{xx}(k) = E(x[n+k]x^*[n])$ since the process is wide-sense stationary and where the superscript asterisk indicates complex conjugation. The input sequence $x[n]$ is white noise if $r_{xx}(k) = \sigma^2 \delta(k)$ for some finite noise variance σ^2 and for the Kroenker delta $\delta(0) = 1$ and $\delta(k) = 0$ if $k \neq 0$. Then passing white noise through the LTI gives the simple relation $R_{yy}(e^{i\omega}) = \sigma^2 |H(e^{i\omega})|^2 = \sigma^2 H(e^{i\omega})H^*(e^{i\omega})$. So we can approximate the output power spectrum $R_{yy}(e^{i\omega})$ by passing unit-variance white noise $x[n]$ through an LTI system whose frequency response $H(e^{i\omega})$ is the "square root" of the power spectrum $R_{yy}(e^{i\omega})$ and then inverse-transforming the frequency response to find the desired LTI filter coefficients $h[n]$. Consider the output power spectrum $R_{yy}(e^{i\omega}) = (5 - 4\cos\omega)^{-1}$. Some manipulation shows that this corresponds to the frequency response $H(e^{i\omega}) = (1 - 2e^{-i\omega})^{-1}$ for a unit-variance white-noise input. Then the desired LTI filter coefficients have the right-sided exponential form $h[n] = 2^n u[n]$. The unit step function $u[n]$ is the binary mapping such that $u[n] = 1$ if $n \geq 0$ and $u[n] = 0$ if $n < 0$.

7. A second-order autoregressive model accurately fits the time-series sunspot data from the year 1770 through the year 1869: $y[n] = 1.318y[n-1] - 0.634y[n-2] + x[n]$ where the zero-mean white-noise sequence $x[n]$ has variance $\sigma^2 = 289.2$. The estimated output power spectrum yields a standardized cumulative periodogram that stays within a 95% confidence band as required if the driving input white noise is Gaussian. The output power spectrum indicates a peak near a value close to the eleven-year periodicity of the sunspot cycle. For details see D. G. Manolakis, V. K. Ingle, and S. M. Kogon, *Statistical and Adaptive Signal Processing*, pp. 448–455, McGraw-Hill, 2000.

8. S. J. Orfanidis, *Introduction to Signal Processing*, p. 712, Prentice Hall, 1996.

9. The LMS algorithm is a simple but random gradient-descent learning algorithm (with a *constant* learning rate) in which the new vector of filter weights w_{k+1} equals the old vector w_k plus a simple and physically available correction term based on the new input signal x_k: $w_{k+1} = w_k + 2\mu e_k x_k$ for learning constant $\mu > 0$. The sequence of random signal vectors $x_1, x_2, \ldots$ is wide-sense stationary with autocorrelation matrix R. The instantaneous error e_k is a scalar random variable that has the linear form of a desired signal minus the actual or observed signal: $e_k = d_k - x_k^T w_k$ for desired input signal d_k that can be deterministic or another wide-sense stationary random sequence. This implies an underlying linear system of the form $y_k = x_k^T w_k$. The deterministic autocorrelation matrix R is not only real symmetric but positive semidefinite since $R = E[x_k x_k^T]$ and since the input signal vectors x_k are real. So R has n non-negative eigenvalues $\lambda_1 \le \lambda_2 \le \cdots \le \lambda_n$. A principal-components analysis shows that the maximum eigenvalue controls the learning rate μ. The LMS algorithm will converge for wide-sense stationary statistics to the global error minimum if the eigenvalue condition $0 < \mu < 1/\lambda_n$ holds. See B. Widrow and S. D. Stearns, *Adaptive Signal Processing*, p. 58, Prentice Hall, 1985. The largest eigenvalue λ_n tends to be large in noisy or highly uncertain environments. Then the eigenvalue condition $0 < \mu < 1/\lambda_n$ gives a small learning rate μ. So the LMS algorithm takes cautious steps as it tries both to estimate and to track the downhill error gradient. The reverse holds in less noisy environments and so a larger learning rate μ then allows larger learning increments and faster convergence to the mean-squared optimal filter weights.

   The statistically more robust *signed* LMS algorithm results by replacing the potentially impulsive error term e_k with its clipped or signed term sgn(e_k) that equals -1 or 1 depending on whether the observed error realization e_k is negative or non-negative. This gives the signed LMS algorithm as $w_{k+1} = w_k + 2\mu \mathrm{sgn}(e_k) x_k$. For details see D. G. Manolakis, V. K. Ingle, and S. M. Kogon, *Statistical and Adaptive Signal Processing*, p. 610, McGraw-Hill, 2000.

10. The LMS algorithm first appeared not in a journal but in a conference proceeding: B. Widrow and M. E. Hoff, "Adaptive Switching Circuits," *IRE Wescon Convention Record Part IV*, pp. 96–104, 1960. Stanford electrical engineer Bernard Widrow explains how he came up with the key idea of the LMS algorithm in 1959 while standing at his blackboard with his Ph.D. student Ted Hoff: "On the blackboard I had written expressions for the error and the square of the error for a linear combiner. In the course of discussion a new idea 'popped up' about differentiating the expression for the error to find the gradient. The true gradient is a vector of partial derivatives of the mean-squared error with respect to the weights. The new idea used the square of a single value of error in place of the mean-squared error. This was differentiated and gave an approximate gradient or a gradient estimate." B. Widrow, "Thinking about Thinking: The Discovery of the LMS Algorithm," *IEEE Signal Processing Magazine*, pp. 100–106, January 2005.

Hence Widrow estimated the unknown but *deterministic* mean-squared error $E[e_k^2]$ at time k with a single available realization of the *random* squared-error random variable e_k^2: $E[e_k^2] \approx e_k^2$. The result estimates the probability-weighted average of infinitely many footprints with the single observed footprint. This LMS approximation gives (trivially) an unbiased estimator of the unknown mean-squared error since taking expectations of both sides gives the same value for all times k when viewing e_k^2 as a random variable. The LMS approximation requires well-behaved signal and noise statistics but yields a remarkably simple and computable gradient estimate: $\nabla_w e_k^2 = -2e_k x_k$ where the sequence of random vectors $x_1, x_2, \ldots$ is wide-sense stationary and where the instantaneous error has the linear form $e_k = d_k - x_k^T w_k$ for (often random) desired input signal d_k and filter weight vector w_k. Inserting this gradient estimate into the general gradient-descent schema $w_{k+1} = w_k - \mu \nabla_w E[e_k^2]$ gives the LMS algorithm in the previous note. Most modern neural-network learning algorithms result when one inserts a similar gradient estimate but for output errors that arise from various types of nonlinear systems.

11. For a discussion of the role that the IEEE First International Conference on Neural Networks (IEEE ICNN-87) in 1987 played in the field of neural networks see chapters 13 and 17 of J. A. Anderson and E. Rosenfeld, eds., *Talking Nets: An Oral History of Neural Networks*, MIT Press, 1998.

12. The simplest form of the least mean absolute deviation (LMAD) algorithm is signed LMS as discussed in note 9 above. More formal versions modify the error term according to the type of symmetric alpha-stable impulsive noise that the learning system encounters (see note 7 to chapter 4 for a discussion of stable distributions and note 11 to chapter 6 for the related use of a Cauchy noise suppressor). For a complete discussion and LMS comparisons see M. Nikias and M. Shao, *Signal Processing with Alpha-Stable Distributions and Applications*, John Wiley and Sons, 1995.

13. Victor Solo and Xuan Kong, *Adaptive Signal Processing Algorithms: Stability and Performance*, p. 3, Prentice Hall, 1995. This application of LMS to fetal electrocardiography grows out of Bernard Widrow's earlier work in B. Widrow, J. R. Glover et al., "Adaptive Noise Cancelling: Principles and Applications," *Proceedings of the IEEE*, vol. 63, pp. 1692–1715, 1975.

14. The term "intelligent signal processing" or ISP stems from the November 1998 special issue on ISP in the *Proceedings of the IEEE*. An expanded version of the special issue appears as the edited volume S. Haykin and B. Kosko, eds., *Intelligent Signal Processing*, Wiley-IEEE Press, 2001.

15. United States patent number 5,539,769, "Adaptive Fuzzy Frequency Hopping System," B. Kosko and P. J. Pacini, issued on 23 July 1996. For information on the history and techniques of spread spectrum communications see R. A. Scholtz, "Notes on Spread Spectrum History," *IEEE Transactions on Communications*, vol. 31, no. 1, pp. 82–84, January 1983; R. Price, "Further Notes and Anecdotes on Spread-Spectrum Origins," *IEEE Transactions on Communications*, vol. 31, no. 1, pp. 85–97, January 1983; M. K. Simon, J. K. Omura, R. A. Scholtz, and B. K. Levitt, *Spread Spectrum Communications Handbook*, McGraw-Hill, 1994; R. L. Peterson, R. E. Ziemer, and D. E. Borth, *Introduction to Spread Spectrum Communications*, Prentice Hall, 1995.

CHAPTER 6: THE ZEN OF NOISE

1. Stochastic resonance occurs only in nonlinear systems: "SR is a phenomenon whereby the addition of a random interference ('noise' as it is almost universally called) can enhance the detection of weak stimuli or enhance the information content of a signal (for example: trains of action potentials or signals generated by a neuronal assembly)." F. Moss, L. M. Ward, and W. G. Sannita, "Stochastic Resonance and Sensory Information Processing: A Tutorial and Review of Applications," *Clinical Neurophysiology*, vol. 115, pp. 267–281, 2004. Both the above quote and the text quote come from page 268 of the article.

The awkward term "stochastic resonance" stems from the 1981 article R. Benzi, A. Sutera, and A. Vulpiani, "The Mechanism of Stochastic Resonance," *Journal of Physics A*, vol. 14, no. 11, pp. 453–457, 1981. Its purported application to explain ice ages appeared the following year in C. Nicolis, "Stochastic Aspects of Climatic Transitions–Response to a Periodic Forcing," *Tellus*, vol. 34, no. 1, pp. 1–9, 1982. But a noise benefit can occur both from nonstochastic interference such as chaos and in nonresonant feedforward systems such as a simple memoryless threshold neuron. The term "noise benefit" is both more accurate and more general given the broad definition of noise as an unwanted signal and with the understanding that desired and undesired signals occur in a nonlinear system. A finer point is that the SR response can be linear in some special but important electrical circuits with periodic input signals but even these effects occur in circuit systems that are themselves nonlinear: D. G. Luchinsky, R. Mannella, P. V. E. McClintock, and N. G. Stocks, "Stochastic Resonance in Electrical Circuits–I: Conventional Stochastic Resonance," *IEEE Transactions on Circuits and Systems–II: Analog and Digital Signal Processing*, vol. 46, no. 9, pp. 1205–1214, September 1999.

2. The stochastic resonance effect in figure 6.1 requires adding white Laplace noise to the thresholded image pixels of the standard baboon test image of signal and image processing. The output threshold has the form $y = \text{sgn}[(x + n) - T]$ for pixel input $x \in [0, 1]$ and additive zero-mean Laplacian white noise n and numerical threshold T. The signum function has the form $\text{sgn}[x] = 1$ if $x \geq 0$ and $\text{sgn}[x] = 0$ if $x < 0$. The simulation used the threshold value $T = 0.04$. Then thresholding the original baboon image gave the faint image in part (a) of figure 6.1. Adding faint Laplace noise with variance $\sigma_n^2 = 0.01$ gave the noisy baboon image in (b). Adding more energetic Laplace noise with $\sigma_n^2 = 0.05$ gave the noisier image

in (c) and adding such noise with $\sigma_n^2 = 0.5$ gave the still noisier and degraded image in (d). The probability density function of the zero-mean Laplace noise n had the form $f(n) = \dfrac{e^{-|n/\beta|}}{2\beta}$ for $\sigma_n^2 = 2\beta^2$. Adding white uniform noise produced the SR effect in the Lena image in chapter 1. Adding infinite-variance Cauchy noise still produces the SR effect in the Lena (and other images) as in S. Mitaim and B. Kosko, "Adaptive Stochastic Resonance in Noisy Neurons Based on Mutual Information," *IEEE Transactions on Neural Networks*, vol. 15, no. 6, pp. 1526–1540, November 2004.

3. Adding a noise "dither" occurs in many areas of signal and image processing to smooth out the perceived effects of digitizing sound or images: "Another method of suppressing contouring effects is to add a small amount of uniformly distributed pseudorandom noise to the luminance samples before quantization. The effect is that in the regions of low-luminance gradients (which are the regions of contours) the input noise causes pixels to go above or below the original decision level, thereby breaking the counters. The average value of the quantized pixels is about the same with and without the additive noise. The amount of dither added should be kept small enough to maintain the spatial resolution but large enough to allow the luminance values to vary randomly about the quantizer decision levels." A. Jain, *Fundamentals of Digital Image Processing*, pp. 120–121, Prentice Hall, 1989. Chapter 5 discussed a similar use of adding noise to smooth out digital quantization in sound signals.

The military allegedly began using dithering in World War II to gently jostle electrical equipment after observing that such equipment worked best aboard flying aircraft: "Airplane bombers used mechanical computers to perform navigation and bomb trajectory calculations. Curiously, these computers (boxes filled with hundreds of gears and cogs) performed more accurately when flying on board the aircraft, and less well on ground." K. C. Pohlmann, *Principles of Digital Audio*, 4th ed., p. 46, McGraw-Hill, 2005. Control engineers also use a related form of noise injection to enhance faint signals: "One method of counteracting the dead zone produced by nonlinear components of friction is to use a dither signal. A dither signal is an alternating signal (usually sinusoidal) that is added electronically to the control signal in the forward path. The dither is introduced to superimpose a vibration on the developed torque in a manner that will tend to incite motion despite a very low signal level." P. H. Lewis and C. Yang, *Basic Control Systems Engineering*, p 373, Prentice Hall, 1997.

Simple forms of threshold stochastic resonance act as noise dither and vice versa: L. Gammaitoni, "Stochastic Resonance and the Dithering Effect in Threshold Physical Systems," *Physical Review E*, vol. 52, no. 4, pp. 4691–4698, November 1995; R. A. Wannamaker, S. P. Lipshitz, and J. Vanderkooy, "Stochastic Resonance as Dithering," *Physical Review E*, vol. 61, no. 1, pp. 233–236, January 2000. Luca Gammaitoni analyzes the SR dithering effect when one adds noise to the original image before the system quantizes or applies a pixel threshold. Choose grayscale pixel value $x \in [0, 1]$ and denote the binary output pixel as $y \in \{0, 1\}$ with threshold $T = 1/2$. Then the dithered quantizer gives $E[Y \mid X = x] = 1 - P(n < T - x) = x$ if and only if the noise is uniform on the interval $(-\frac{1}{2}, \frac{1}{2})$. This can be only a partial explanation of the SR effect because the effect persists for not only nonuniform dithering noise but even for infinite-variance dithering noise. See the preceding note for a reference to such an SR effect in a Cauchy-dithered "Lena" image.

We further note that the literature of neural networks has long since recognized the related advantages of deliberately adding noise to improve learning and even to approximate the effect of adding second-derivative "regularization" terms to a network's global squared-error performance measure: K. Matsuoka, "Noise Injection into Inputs in Back-Propagation Learning," *IEEE Transactions on Systems, Man, and Cybernetics*, vol. 22, no. 3, pp. 436–440, May 1992; C. M. Bishop, "Training with Noise Is Equivalent to Tikhonov Regularization," *Neural Computation*, vol. 7, no. 1, pp. 108–116, 1995; Y. Grandvalet, S. Canu, and S. Boucheron, "Noise Injection: Theoretical Prospects," *Neural Computation*, vol. 9, no. 5, pp. 1093–1108, 1997; M. Burger and A. Neubauer, "Analysis of Tikhonov Regularization for Function Approximation by Neural Networks," *Neural Networks*, vol. 16, no. 1, pp. 79–90, January 2003.

4. Stochastic resonance and noise benefits also extend to spatiotemporal pattern formation and other effects in wave propagation: "While single disturbances can only produce temporary structures, since they are eventually convected out of the (finite) system, continuous noise will develop permanent *noise sustained structures*," M. Loecher, *Noise Sustained Patterns*, World Scientific Lecture Notes in Physics, vol. 70, World Scientific, 2003. For an early example see R. J. Deissler, "External Noise and the Origin and Dynamics of Structure in Convectively Unstable Systems," *Journal of Statistical Physics*, vol. 54, no. 5, pp. 1459–1488, March 1989. A more recent review is P. Jung, A. Cornell-Bell, F. Moss, S. Kadar, J. Wang, and K. Showalter, "Noise Sustained Waves in Subexcitable Media: From Chemical Waves to Brain Waves," *Chaos*, vol. 8, no. 3, pp. 567–575, September 1998.

Additive noise also appears to stabilize chaotic or aperiodic patterns in coupled-oscillator models of the olfactory system: "Chaotic solutions having biological verisimilitude are robustly stabilized by introducing low-level, additive noise from a random number generator at two biologically determined points: rectified, spatially incoherent noise on each receptor input line, and spatially coherent noise to the anterior olfactory nucleus, a global control point receiving centrifugal inputs from various parts of the forebrain." W. J. Freeman, H. J. Chang, B. C. Burke, P. A. Rose, and J. Badler, "Taming Chaos: Stabilization of Aperiodic Attractors by Noise," *IEEE Transactions on Circuits and Systems*, vol. 44, no. 10, pp. 989–996, October 1997.

5. The holy grail of research in biological stochastic research is to demonstrate an internal or endogenous noise benefit at work in living brain tissue: "To identify 'internal' noise in the central nervous system and to incorporate in the SR theory any of the possible sources of neuronal noise (ion channel or synaptic noise, noise built into the stimulus, light, eye micro-movements, etc.) remains a major and still unresolved problem. Once 'internal' noise originating from any source is defined, also crucial is how to change it in the brain under experimental control or to monitor and relate to function its spontaneous or state-related fluctuations." Frank Moss, Lawrence M. Ward, and Walter G. Sannita, "Stochastic Resonance and Sensory Information Processing: A Tutorial and Review of Applications," *Clinical Neurophysiology*, vol. 115, pp. 267–281, 2004.

So far the best evidence of a *behavioral* SR effect from naturally occurring electrical noise appears to be that of the river paddlefish's response to the swarms of prey plankton and the faint electrical fields they emit as they swim and eat: "Stochastic resonance requires an external source of electrical noise in order to function. A swarm of plankton, for example *Daphnia*, can provide the required noise. We hypothesize that juvenile paddlefish can detect and attack single *Daphnia* as outliers in the vicinity of the swarm by using noise from the swarm itself." J. A. Freund, L. Schimansky-Geier, B. Beisner, A. Neiman, D. F. Russell, T. Yakusheva, and

F. Moss, "Behavioral Stochastic Resonance: How the Noise from a *Daphnia* Swarm Enhances Individual Prey Capture by Juvenile Paddlefish," *Journal of Theoretical Biology*, vol. 214, pp. 71–83, 2002.

6. Scientists have observed some form of a stochastic resonance noise benefit in several models of pulsed or spiking neurons: "Sensory neurons transform signals from the environment into trains of spikes that propagate to other structures in the nervous system. Since internal and external noises are ubiquitous and unavoidable, many studies have investigated their effect on signal transmission by sensory neurons. These have shown that noise of appropriate amplitude linearizes the response of neurons, leads to stochastic resonance, and maximizes input-output correlation (power norm), transinformation, and coherence." T. Shimokawa, A. Rogel, K. Pakdaman, and S. Sato, "Stochastic Resonance and Spike-Timing Precision in an Ensemble of Leaky Integrate and Fire Neuron Models," *Physical Review E*, vol. 59, no. 3, pp. 3461–3470, March 1999. One of the earliest papers to explore SR noise benefits in a dedicated neuron model was A. Bulsara, E. W. Jacobs, T. Zhou, F. Moss, and L. Kiss, "Stochastic Resonance in a Single Neuron Model: Theory and Analog Simulation," *Journal of Theoretical Biology*, vol. 52, no. 4, pp. 531–555, October 1991.

Related SR findings in biologically plausible neuron models and direct animal experiments include J. K. Douglass, L. Wilkens, E. Pantazelou, and F. Moss, "Noise Enhancement of Information Transfer in Crayfish Mechanoreceptors by Stochastic Resonance," *Nature*, vol. 365, pp. 337–340, 23 September 1993; J. E. Levin, and J. P. Miller, "Broadband Neural Encoding in the Cricket Cercal Sensory System Enhanced by Stochastic Resonance," *Nature*, vol. 380, pp. 165–168, 14 March 1996; J. J. Collins, T. T. Imhoff, and P. Grigg, "Noise-Enhanced Information Transmission in Rat

SA1 Cutaneous Mechanoreceptors via Aperiodic Stochastic Resonance," *Journal of Neurophysiology*, vol. 76, no. 1, pp. 642–645, July 1996; W. C. Stacey and D. M. Durand, "Stochastic Resonance Improves Signal Detection in Hippocampal CA1 Neurons," *Journal of Neurophysiology*, vol. 83, no. 3, pp. 1394–1402, March 2000; M. Yoshida, H. Hayashi, K. Tateno, and S. Ishizuka, "Stochastic Resonance in the Hippocampal CA3–CA1 Model: A Possible Memory Recall Mechanism," *Neural Networks*, vol. 15, no. 10, pp. 1171–1183, December 2002.

7. The strongest version of the forbidden interval theorem holds for simple threshold neurons (or other thresholdlike systems with numerical threshold T) that receive noisy bipolar subthreshold signals $-A$ and A as a random Bernoulli sequence: Stochastic resonance occurs in the sense of increasing the neuron's input-output Shannon mutual information $I(S, Y)$ given the dispersion of the independent additive noise if and only if the noise mean $E(n)$ does not lie in the "forbidden interval" $(T-A, T+A)$ where $-A < A < T$ for signal amplitude A. The input signal S_t is a Bernoulli random sequence such that S_t equals either $-A$ or A at time t with arbitrary success probability p so long as $0 < p < 1$. The noisy thresholded output has the form $Y_t = \mathrm{sgn}(S_t + n_t - T)$ for independent noise process n_t. The mutual information $I(S, Y)$ is non-negative because conditioning reduces entropy: $I(S, Y) = H(Y) - H(Y|S) \geq 0$ since $I(S, Y) = \sum_s \sum_y p(s, y) \log_2 \dfrac{p(s, y)}{p(s)p(y)}$. The theorem holds for *any* noise process n with finite variance. It also holds for any infinite-variance noise type from the class of stable distributions—but then the noise location parameter replaces the noise mean just as the dispersion parameter replaces the variance. The sufficient or if-part of the theorem first appeared in B. Kosko and S. Mitaim, "Stochastic Resonance in Noisy Threshold Neurons," *Neural Networks*,

vol. 16, no. 6, pp. 755–761, June 2003. The necessary or then-part of the theorem first appeared in B. Kosko and S. Mitaim, "Robust Stochastic Resonance for Simple Threshold Neurons," *Physical Review E*, vol. 70, pp. 031911-1–031911-10, 27 September 2004. The only-if part holds only in the sense that the random threshold neuron performs better bit-wise without noise than with it when the interval condition fails. The theorem does not guarantee that the noise benefit in terms of an increased bit count will be substantial or even detectable in any given instance. The theorem guarantees only that some noise benefit will occur.

The proof technique or strategy is to show that what goes down can go up. The proof first assumes that the non-negative mutual information $I(S, Y)$ is positive and thus that there exists some correlation between the neuron system's random input signal S and its random output Y. Then the proof shows that the mutual information goes to zero (and thus that input S and output Y are statistically independent) as the noise variance or dispersion σ goes to zero: $\lim_{\sigma \to 0} I(S, Y) = 0$. So there must be some noise benefit because then the mutual information or bit count must increase if the noise dispersion increases from zero. This proof technique generalizes to many other stochastic nonlinear systems because it replaces the difficult task of showing that the qualitative property of an SR noise benefit occurs with the comparatively simpler task of showing that a limit equals zero. Yet establishing this zero limit can involve a fair amount of analysis for more complex neuronal dynamics and for any such result cast in terms of the mean-square stochastic calculus.

8. So-called suprathreshold stochastic resonance arises only for two or more threshold neurons if the additive noise is uniform: "It [suprathreshold SR] differs from SR in a single element [threshold neuron] in a number of ways. First, it only occurs in arrays

[networks] of two or more elements. . . . Second, it is at its most pronounced when the deterministic threshold crossings are maximized—this is achieved by setting the threshold levels to coincide with the DC signal component. . . . Third, supra-threshold SR can occur for arbitrary binary signal strengths. It does not require that the signal be weak [subthreshold]." Nigel G. Stocks, "Supra-threshold Stochastic Resonance: An Exact Result for Uniformly Distributed Signal and Noise," *Physics Letters A*, vol. 279, pp. 308–312, 5 February 2001.

9. Noise appears to improve the suboptimal performance of many classical statistical decision-theoretic systems even though true optimal performance occurs in the ideal case of no noise interference: "A stochastic resonance effect, under the form of a noise-improved performance, is shown feasible for a whole range of optimal detection strategies, including Bayesian, minimum error-probability, Neyman-Pearson, and minimax detectors." D. Rousseau and F. Chapeau-Blondeau, "Stochastic Resonance and Improvement by Noise in Optimal Detection Strategies," *Digital Signal Processing*, vol. 15, pp. 19–32, 2005. For more details on improving a Bayesian estimator in the presence of corrupting phase noise see also F. Chapeau-Blondeau and D. Rousseau, "Noise-enhanced Performance for an Optimal Bayesian Estimator," *IEEE Transactions on Signal Processing*, vol. 52, no. 5, pp. 1327–1334, May 2004. For related noise benefits in suboptimal signal systems see S. Kay, "Can Detectability Be Improved by Adding Noise?" *IEEE Signal Processing Letters*, vol. 7, no. 1, pp. 8–10, January 2000.

10. Two forbidden interval theorems for several standard models of *spiking* retinal and sensory neurons appear in A. Patel and B. Kosko, "Stochastic Resonance in Noisy Spiking Retinal and Sensory Neuron Models," *Neural Networks*, vol. 18, pp. 467–478, July

2005. The first theorem gives necessary and sufficient condition for a stochastic resonance (SR) noise benefit for subthreshold signals in a standard family of Poisson spiking models of retinal neurons: SR holds if and only if a sum of noise means or location parameters (for stable infinite-variance noise) falls outside the forbidden interval of values. The only-if part or necessary condition holds only in the sense that the system performs better without noise than with it when the interval condition fails. The second theorem gives a similar forbidden interval sufficient condition for the SR effect for several types of spiking sensory neurons that include the Fitzhugh-Nagumo neuron and the leaky integrate-and-fire neuron as well as the reduced Type I neuron model if the additive noise is Gaussian white noise. Neither Gaussian noise nor the forbidden interval condition appears to be necessary for the SR effect in general for these and related sensory neurons.

11. Adaptive stochastic resonance involves gradually adapting the standard deviation or dispersion of the added noise in a slow stochastic gradient ascent. The first such algorithms appeared in S. Mitaim and B. Kosko, "Adaptive Stochastic Resonance," *IEEE Proceedings*, vol. 86, no. 11, pp. 2152–2183, November 1998. The paper produced a general and statistically robust learning algorithm for several types of noise as well as for additive chaos. The algorithms slowly vary the input noise standard deviation σ_k at time k to maximize the out spectral signal-to-noise ratio (SNR) or other system performance measure. The algorithms have the robustified form $\sigma_{k+1} = \sigma_k + \mu_k \phi \left(\dfrac{\partial SNR}{\partial \sigma} \right)$ for decreasing learning coefficients μ_k and for Cauchy robustifier $\phi(z) = \dfrac{2z}{1+z^2}$ from the theory of robust statistics (see P. J. Huber, *Robust Statistics*, Wiley, 1981). The robustifier cancels Cauchy-like noise spikes in the learning process that arise from the ratio structure of the key learning term. A related algorithm applies to SNR optimization in the presence of added infinite-variance stable noise: B. Kosko and S. Mitaim, "Robust Stochastic Resonance: Signal Detection and Adaptation in Impulsive Noise," *Physical Review E*, vol. 64, pp. 051110/1–051110/11, 22 October 2001. A simpler robustified algorithm results if the system performance measure is Shannon mutual information $I(S, Y) = H(S) - H(S|Y)$ and finds the optimal SR noise level in both sigmoidal and Gaussian (radial-basis) neurons: S. Mitaim and B. Kosko, "Adaptive Stochastic Resonance in Noisy Neurons Based on Mutual Information," *IEEE Transactions on Neural Networks*, vol. 15, no. 6, pp. 1526–1540, December 2004. The resulting algorithm uses only the sign or signum of the learning increment $\frac{\partial I}{\partial \sigma}$: $\sigma_{k+1} = \sigma_k + \mu_k \, \mathrm{sgn}\left(\frac{\partial I}{\partial \sigma}\right)$. The signum operator acts as crude robustifier because it maps all learning values or spikes into either the number −1 or 1 depending on whether they are negative or non-negative. Further assumptions simplify the learning term $\frac{\partial I}{\partial \sigma}$ depending on the underlying system dynamics.

12. The arrangement of the carbon atoms in a carbon nanotube determines whether the nanotube acts as a metal or semiconductor as well as many other of its properties: "Single-walled nanotubes (SWNTs) consist of a single graphite sheet seamlessly wrapped into a cylindrical tube. Multi-walled nanotubes (MWNTs) comprise an array of such nanotubes that are concentrically nested like rings of a tree trunk. Despite structural similarity to a single sheet of graphite, which is a semiconductor with zero band gap, SWNTs may be either metallic or semiconducting depending on the sheet direction about which the graphite sheet is rolled to form a nanotube cylinder. This direction in the graphite sheet plane and the nanotube diameter are obtainable from a pair of integers (n, m) that denote the nanotube type. Depending on the appearance of a belt

of carbon bonds around the nanotube diameter, the nanotube is either of the armchair ($n=m$), zigzag ($n=0$ or $m=0$), or chiral (any other n and m) variety. All armchair SWNTs are metals. Those with $n-m=3k$, where k is a nonzero integer, are semiconductors with a tiny band gap. And all others are semiconductors with a band gap that inversely depends on the nanotube diameter. . . . Because of the nearly one-dimensional electronic structure, electronic transport in metallic SWNTs and MWNTs occurs ballistically (i.e., without scattering) over long nanotube lengths, enabling them to carry high currents with essentially no heating. Phonons also propagate easily along the nanotube: The measured room temperature conductivity for an individual MWNT (>3000 W/mK) is greater than that of natural diamond and the basal plane of graphite (both 2000 W/mK). . . . Small-diameter SWNTs are quite stiff and exceptionally strong, meaning that they have a high Young's modulus and high tensile strength." R. H. Baughman, A. A. Zakhidov, and W. A. de Heer, "Carbon Nanotubes–the Route Toward Applications," *Science*, vol. 297, pp. 787–792, 2 August 2002. Carbon nanotubes further permit various forms of superconductivity and supercurrents (dissipationless currents) as in P. Jarillo-Herrero, J. A. van Dam, and L. P. Kouwenhoven, "Quantum Supercurrent Transistors in Carbon Nanotubes," *Nature*, vol. 439, pp. 953–956, 23 February 2006.

13. A quantum dot is a type of artificial atom or semiconductor nanocrystal that traps and manipulates electrons: "The reduction in dimensionality produced by confining electrons (or holes) to a thin semiconductor layer leads to a dramatic change in their behavior. This principle can be developed by further reducing the dimensionality of the electron's environment from a two-dimensional quantum well to a one-dimensional quantum wire and eventually to a zero-dimensional quantum dot." P. Harrison,

Quantum Wells, Wires, and Dots: Theoretical and Computational Physics of Semiconductor Nanostructures, p. 243, Wiley, 2005.

14. Peter Burke and his research team at the University of California at Irvine showed that a single-walled carbon nanotube transistor could switch current on and off at about a tenth of a nanosecond (in a *cold* environment of 4 degrees kelvin): "We present the first demonstration of single-walled carbon nanotube transistor operation at microwave frequencies. To measure the source-drain ac current and voltage at microwave frequencies, we construct a resonant LC impedance-matching circuit at 2.6 GHz." S. Li, Z. Yu, S. Yeng, W. C. Tang, and P. J. Burke, "Carbon Nanotube Transistor Operation at 2.6 GHz," *Nano Letters*, vol. 4, no. 4, pp. 753–756, March 2004.

15. Three types of electrical whitelike noise improved three different measures of a nanotube transistor's performance: "Experiments confirm that small amounts of noise help a nanotube transistor detect noisy subthreshold electrical signals. Gaussian, uniform, and impulsive (Cauchy) noise produced this feedforward stochastic-resonance effect by increasing both the nanotube system's mutual information and its input-output correlation. The noise corrupted a synchronous or Bernoulli random digital sequence that fed into the threshold-like nanotube transistor and produced a Bernoulli sequence." I.Y. Lee, X. Liu, B. Kosko, and C. Zhou, "Nanosignal Processing: Stochastic Resonance in Carbon Nanotubes that Detect Subthreshold Signals," *Nano Letters*, vol. 3, no. 12, pp. 1683–1686, December 2003.

16. Superconducting quantum interference devices or SQUIDs are supersensitive detectors of faint magnetic fields. The first experimental demonstration of an SR noise benefit in a SQUID

appeared in 1995: "In our experiment a magnetic flux $\Phi_e(t)$ was imposed on the [radio-frequency SQUID] loop and the total flux $\Phi_t(t)$ through the loop monitored by measuring [the loop current] i. The strength of the stochastic component of the applied flux was varied, and by plotting the SNR (power in the signal bandwidth divided by the average noise power in the same bandwidth) as a function of the [white Gaussian] noise strength, the stochastic resonance effect was quantified." A. D. Hibbs, A. L. Singsaas, E. W. Jacobs, A. R. Bulsara, J. J. Bekkedahl, and F. Moss, "Stochastic Resonance in a Superconducting Loop with a Josephson Junction," *Journal of Applied Physics*, vol. 77, no. 6, pp. 2582–2590, 15 March 1995. For a more formal treatment of SQUID SR effects see M. E. Inchiosa, A. R. Bulsara, and L. Gammaitoni, "Higher-order Resonant Behavior in Asymmetric Nonlinear Stochastic Systems," *Physical Review E*, vol. 55, no. 4, pp. 4049–4056, April 1997.

17. There have been some preliminary efforts to find an SR noise benefit in quantum information systems based on statistical entanglement of quantum states: "Entanglement is present in the system at all times and its exact amount is a nonmonotonic function of the intensity of the external incoherent driving [noise]. This behavior resembles the phenomenon of stochastic resonance, where the response of a nonlinear system to weak periodic driving can be enhanced when supplemented with a noisy field of certain optimal intensity. In our case, the amount of entanglement of the two light fields is maximized for an optimal value of the external noisy field." M. B. Plenio and S. F. Huelga, "Entangled Light from White Noise," *Physical Review Letters*, vol. 88, no. 19, pp. 197901-1–197901-7, 13 May 2002. See also X. X. Yi, C. S. Yu, L. Zhou, and H. S. Song, "Noise-Assisted Preparation of Entangled Atoms," *Physical Review A*, vol. 68, pp. 052304-1–052304-4, November 2003; J.-B. Xu and S.-B. Li, "Control of the Entanglement of Two Atoms in

an Optical Cavity via White Noise," *New Journal of Physics*, vol. 7, no. 1, pp. 72–89, 2005.

18. J. V. Hernandez, E. R. Kay, and D. A. Leigh, "A Reversible Synthetic Rotary Molecular Motor," *Science*, vol. 306, pp. 1532–1537, 26 November 2004.

19. Several molecular motors appear to use thermal noise fluctuations as a driving force: "Consider RNA polymerase, an enzyme that moves along DNA to produce a newly synthesized RNA strand– a process called transcription. Although it has not yet been proven unequivocally, evidence suggests that during transcription the polymerase moves by extracting energy from the thermal bath and uses bond hydrolysis to ensure that only 'forward' fluctuations are captured. That is, the enzyme rectifies thermal fluctuations to directed motion." C. Bustamante, J. Liphardt, and F. Ritort, "The Nonequilibrium Thermodynamics of Small Systems," *Physics Today*, pp. 43–48, July 2005. The quote in the text comes from the same page of this article.

Biophysicist Dean Astumian describes how the protein kinesin uses rectified Brownian motion as it "walks" along microtubules: "Another important example is kinesin, a molecular forklift that transports proteins within the cell. Kinesin consists of two loosely attached domains and moves along a track called a microtubule, made of many individual molecules of the protein tubulin, each about 10 nanometers long. The electric potential between the kinesin and the microtubule usually has a sawtooth pattern, with energy barriers preventing the motion of kinesin from one tubulin molecule to the next. In the Brownian model, hydrolysis of an ATP molecule changes this potential to a flat shape and allows random collisions to jostle the kinesin. Release of the hydrolysis products returns the potential to the sawtooth shape, which, depending on how far the kinesin has drifted, can push the molecule forward." R. D. Astumian, "Making Molecules into Motors," *Scientific American*, pp. 56–64, July 2001.

An important paper on how kinesin motors can arise from rectified Brownian diffusions is R. F. Fox and M. H. Choi, "Rectified Brownian Motion and Kinesin Motion Along Microtubules," *Physical Review E*, vol. 63, pp. 051901–051913, 2001. For related technical details see R. D. Astumian and I. Derenyi, "A Chemically Reversible Brownian Motor: Application to Kinesin and Ncd," *Biophysical Journal*, vol. 77, no. 2, pp. 993–1002, August 1999; R. D. Astumian, "Biasing the Random Walk of a Molecular Motor," *Journal of Physics: Condensed Matter*, vol. 17, pp. S3753–S3766, November 2005.

20. Several papers explain how Brownian motors use asymmetric potentials to achieve direct motion without violating the second law of thermodynamics: M. Bier, "Brownian Ratchets in Physics and Biology," *Contemporary Physics*, vol. 38, no. 6, pp. 371–379, 1997; F. Julicher, A. Adjari, and J. Prost, "Modeling Molecular Motors," *Reviews of Modern Physics*, vol. 69, no. 4, pp. 1269–1281, October 1997; M. O. Magnasco and G. Stolovitzky, "Feynman's Ratchet and Pawl," *Journal of Statistical Physics*, vol. 93, no. 3, pp. 615–632, 1998; P. Reimann and P. Hänggi, "Introduction to the Physics of Brownian Motors," *Applied Physics A*, vol. 75, pp. 169–178, 2002; H. Linke, M. T. Downtown, and M. J. Zuckermann, "Performance Characteristics of Brownian Motors," *Chaos*, vol. 15, pp. 026111-1–026111-11, 2005.

The Bustamante paper in the previous note gives a good introduction to modern fluctuation theorems and how they apply to Brownian motors. Nanotech research has also viewed Brownian motors from a chemical perspective: W. F. Paxton et al., "Catalytic Nanomotors: Autonomous Movement of Striped

Nanorods," *Journal of the American Chemical Society*, vol. 126, pp. 13424–13431, 2004; V. Krishnamurthy and S. Chung, "Brownian Dynamics Simulation for Modeling Ion Permeation across Bionanotubes," *IEEE Transactions on Nanobioscience*, vol. 4, no. 1, pp. 102–111, March 2005.

Page numbers in *italics* refer to illustration captions.